JITTER
IN DIGITAL
TRANSMISSION SYSTEMS

The Artech House Telecommunication Library

Vinton G. Cerf, *Series Editor*

JITTER

IN DIGITAL
TRANSMISSION SYSTEMS

PATRICK R. TRISCHITTA
EVE L. VARMA

ARTECH HOUSE

Library of Congress Cataloging-in-Publication Data

Trischitta, Patrick R., 1958-
    Jitter in Digital Transmission Systems.

    Bibliography: p.
    Includes index.
    1. Digital communications.   2. Signal processing — Digital
    techniques.   3. Optical communications.   I. Varma, Eve L.   II. Title
TK5103.7.T74   1989   621.38′0414   88-8112
ISBN   0-89006-248-X

International Standard Book Number: 0-89006-248-X
Library of Congress Catalog Card Number: 88-8112

   10   9   8   7   6   5   4   3   2

To my wife Elaine Trischitta
and my parents, Patrick and Louise Trischitta

To my family, especially my children
Anjali and Sarita Varma,
and in memory of my father, Ivan D. London

Contents

Preface

Throughout the world, telecommunication networks are employing more and more digital technology to provide higher quality, less expensive, and more varied services to an increasing number of users. As a result, digital transmission systems continue to be deployed at a rapid pace, to add capacity and to replace analog facilities in telecommunication networks that span continents and cross oceans. This evolution began in the early 1960s when AT&T introduced the first digital transmission system to increase the capacity of intracity cable routes. Today, digital transmission systems, particularly, fiber optic transmission systems, have dramatically transformed telecommunication networks worldwide into networks where voice, data, and video are transported together between large digital switching centers as digitally encoded pulses of light propagating through optical fibers.

Despite the almost universal appeal of digital transmission systems, there exists a transmission anomaly that must be considered in designing, developing, deploying, interconnecting, and maintaining every digital system and network. This transmission anomaly is called jitter and has been studied extensively beginning in the 1950s when the first digital systems were conceived. Surprisingly, although many good technical papers have analyzed and discussed jitter, a unified and comprehensive treatment does not exist.

Our purpose in writing this book is to provide a unified and comprehensive treatise on all aspects of jitter. To accomplish this, we have merged the results of over thirty years of investigation into the sources, the accumulation, and the effect of jitter in digital transmission systems. This book grew out of our own work on jitter done in the laboratory, in the field, and in the standards arena over the past eight years. Theory has been blended with practice to show how one applies theoretical principles to design and characterize systems and networks. Results of measurements and simulations on fiber optic transmission systems provide examples and verify the analytical expressions derived. We have concentrated on fiber optic transmission systems because they are now the preferred method for transporting huge quantities of information over long distances, and they will continue to be deployed, with increasing capacity and complexity, into the evolving worldwide digital network. However, the jitter topics discussed and the results obtained can also be applied to coaxial cable, satellite, or digital radio transmission systems.

The intended audience of this book includes practicing professionals involved in developing, designing, system engineering, deploying, and maintaining digital transmission systems and networks; it should prove to be a valuable reference because of its complete and rigorous treatment of the subject. This book may also be valuable to second year graduate students of digital communications. Issues relating to timing and synchronization are usually treated briefly or omitted in first year graduate texts on digital communications. This book may fill in those gaps. The study of jitter requires a thorough background in both digital communications and stochastic processes. The mathematical details presented here may become tedious at times but is necessary for a complete treatment.

The book contains nine chapters.

Chapter 1 introduces jitter in digital transmission systems and describes a fiber optic transmission system.

Chapter 2 describes line regenerators and timing extraction. The jitter introduced and transferred by a regenerator is then modeled as a linear system.

Chapter 3 discusses the accumulation of jitter for cascaded regenerators.

Chapter 4 analyzes the effect jitter has on transmission quality. The concept of jitter tolerance and its relationship with alignment jitter is discussed.

Chapter 5 describes the concept of time-digital multiplexing and the various approaches for characterizing multiplex jitter.

Chapter 6 discusses jitter tolerance and jitter transfer in digital multiplexes.

Chapter 7 analyzes jitter accumulation in complex digital networks, including the implications of SONET.

Chapter 8 by Karen E. Plonty, models and measures slowly-varying jitter called wander in fiber optic transmission systems.

Chapter 9 discusses the current status of network jitter standards and control strategies developed within CCITT and ANSI Technical Subcommittee T1X1.

We first wish to acknowledge the support of our management at AT&T Bell Laboratories, especially Judy Page. Her suggestion brought us together for this book and has resulted in a book that is more comprehensive than originally envisioned.

We would like to thank Howard Altman, Samia Bahsoun, John Blake, Christos Chamzas, Yin-Wu Chen, John Ellson, Jon Hill, and Rob Nunn for carefully reviewing the manuscript and for providing many valuable comments. We would also like to thank and acknowledge the help and support of the many professionals worldwide that we have worked with over the years in our study of jitter in digital transmission systems.

Chapter 1
Introduction to Jitter in Digital Transmission Systems

1.1 DEFINING JITTER

In an ideal digital transmission system, the pulses of the digital pulse stream would arrive at times that are integer multiples of the pulse repetition period T. However, in real systems, pulses arrive at times that differ from integer multiples of T. We call this unwanted pulse position modulation of the pulse stream *jitter*. To illustrate jitter, we show in Figure 1.1a a train of unit impulses initially spaced equally in time; after either transmission or signal processing these unit impulses become spaced slightly irregularly in time (Figure 1.1b). The time deviations from integer multiples of T form a discrete time, continuous amplitude sequence $e[nT]$. This sequence is our fundamental description of jitter; $e[nT]$ has dimensions of time in amplitude at integer multiples of T and is positive for a given pulse that arrives earlier than time nT. Throughout this book we use square brackets to denote a discrete time function and parenthesis to denote a continuous time function. We often convert $e[nT]$ to units of degrees by defining T to equal 360 degrees. Therefore, a jitter amplitude of T seconds is 360 degrees or, equivalently, 1 Unit Interval (UI) of jitter.

In digital transmission systems $e[nT]$ is a random function of time called a stochastic process. We apply the mathematics of stochastic processes, with notation and definitions described in [1.1 and 1.2], to quantify and analyze $e[nT]$. We calculate and measure such statistical quantities as the mean, the root mean squared (RMS), and the power spectrum of $e[nT]$. In the following chapters we will study how $e[nT]$ is generated, how it accumulates, and what effect it has on digital transmission systems, particularly fiber optic transmission systems.

The primary sources of jitter in digital transmission systems are regenerators and multiplexes. Digital transmission systems use regenerators to transport information distances longer than a single span of the transmission medium [1.3]. A regenerator receives the incoming pulse stream and transmits a new regenerated pulse

DEFINITION OF JITTER

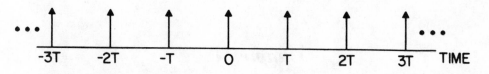

(a) IMPULSES SPACED EQUALLY IN TIME (JITTER FREE SIGNAL)

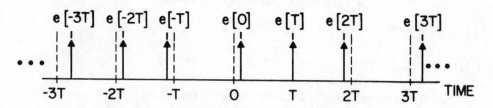

(b) IMPULSES SPACED IRREGULARLY IN TIME
(JITTERED SIGNAL)

Fig. 1.1 Illustration of Jitter

stream that resembles the original as closely as possible. To regenerate the signal, timing information must be known so that the regenerated pulse stream can be transmitted with the proper intervals between pulses. The most commonly used regenerators in digital transmission systems are self-timed regenerators that extract timing information directly from the incoming pulse stream using a timing extraction circuit [1.4]. Because the timing extraction process is imperfect, the transmitted pulse stream is not an exact replica of the original pulse stream but contains the unwanted pulse position modulation we call jitter.

The condition of the received pulse stream as it arrives at the receiver of the regenerator also adds to imperfect timing extraction and regeneration. Corrupted by additive noise from the transmission medium and receiver, the received pulses also have pulse shapes that spill over into adjacent time slots, resulting in intersymbol interference. Thus, the imperfect timing extraction circuit must extract a timing signal from a noisy, dispersed pulse stream. This will result in regenerator output that is inherently jittered. In Chapter 2, we study in detail the jitter introduced by self-timed regenerators. We discuss how a timing signal can be extracted from the received pulse stream and why the timing signal is inherently jittery. We then derive a linear jitter model for the self-timed regenerator.

Since regenerators are cascaded to form digital transmission systems, an accumulation of jitter results from the cascaded regeneration processes. In Chapter 3, we derive jitter accumulation models that assume either non-identical, identical or statistically distributed cascaded regenerators and compare model calculations with results of jitter accumulation measurements made on cascaded fiber optic regenerators.

Accumulated jitter will degrade digital system transmission quality by impairing the regenerator's ability to make correct bit decisions. The effect jitter has on the decision making process inside the regenerator is studied in Chapter 4. In Chapter 4 we define alignment jitter as the difference between the jitter on the received signal to be sampled and the jitter on the extracted clock doing the sampling. We show that it is alignment jitter that affects the decision making process inside the regenerator, and if not controlled, can cause decision errors.

A digital transmission system usually combines several lower rate pulse streams into a higher rate pulse stream using time division multiplexing. To accomplish time division multiplexing, the lower speed pulse streams must be synchronized to a commom rate. There are three synchronization schemes in use today: slip buffering, bit justification, and pointer processing. For example, the commonly used positive bit justification scheme [1.5] is based on having the rate of the outgoing pulse stream of the multiplex higher than the maximum instantaneous sum of the incoming rates. The extra time slots are filled by justifying ("stuffing") extra pulses into each incoming lower speed signal until its rate is raised to that of a local common clock. The justified pulses are inserted at fixed locations in the multiplex frame format so that they may be identified and removed at the demultiplexing terminal. Jitter appears on the lower rate tributaries, generated by both the justification process and the *waiting time* for justification opportunities in the frame format [1.6, 1.7]. We will study the jitter introduced by digital multiplexes in Chapter 5 and how digital multiplexes tolerate and transfer jitter in Chapter 6.

Since digital transmission systems are part of worldwide networks, jitter accumulation in complex digital networks is explored in Chapters 7 and 8. This includes jitter that varies extremely slowly, often with periods of one day to one year. This slowly-varying jitter, called *wander,* can be large (e.g. tens of UIs). Wander, its sources, how it accumulates, and an estimate of its magnitude in fiber optic transmission systems, is discussed in Chapter 8.

As will become evident throughout this book, jitter is an anomaly in digital systems that affects the design of components as well as the design of digital networks. As a result, there has been much effort to standardize definitions, measurement procedures, and system interface requirements. Chapter 9 examines and compares jitter standards and control strategies within two key standards arenas. Furthermore, recent worldwide activities to standardize a *synchronous optical network* (SONET) and a *synchronous digital hierarchy* (SDH) may greatly increase the importance of jitter standards in assuring high quality telecommunications network performance [1.8, 1.9–1.12].

To begin our treatment of jitter, we will briefly review the basic elements of a fiber optic transmission system. In later chapters we will apply our understanding of jitter to this important type of digital transmission system. We have chosen to concentrate on fiber optic transmission systems instead of coaxial, satellite, or digital radio systems, because of fiber's rapid rise to prominence as the preferred method of transmission. Fiber optic transmission systems will continue to be deployed to increase the capacity and complexity of the evolving worldwide digital network. However, the jitter topics discussed and the results obtained can usually be generalized to other types of digital transmission systems and networks.

1.2 FIBER OPTIC TRANSMISSION SYSTEM DESCRIPTION

We will briefly describe elements of a fiber optic transmission system. Since our purpose is to study jitter in digital systems and to apply our results to fiber optic transmission systems, we will not go in detail into the physics of the various fiber optic system components (e.g. lasers, receivers, fiber). Even though fiber optic communication is a new field, many excellent books on the subject are available [1.13–1.23]. These can provide the details of the device and system technology we have omitted so that we can concentrate instead on the study of jitter in these systems.

Key US and international standards have been involved in an effort to establish a synchronous digital hierarchy. The US *synchronous optical network* (SONET) standards provide an optical digital hierarchy extending to high bit rates that flexibly transport different capacity signals [1.8, 1.9]. Compatibility between different manufacturers' fiber optic transmission systems is an objective. The development of SONET standards in the US led to interest and research by other administrations that resulted in the CCITT studies on SDH. The forthcoming CCITT SDH Recommendations are technically consistent with the SONET rates and formats standard but allow for different administrative options. Although not part of SONET or SDH, it is interesting to note the TAT-8 Undersea Lightwave System [1.18] combines systems from three suppliers that meet in an underwater "wet connection." For this system, the line rate, line code, and frame format were standardized by all undersea system manufacturers. However, the high-speed line aspects, both logical and physical, of current asynchronous fiber optic systems are manufacturer specific.

Figure 1.2 shows the basic block diagram of a fiber optic transmission system. M electrical data signals with well defined and standardized rate, line code, and pulse shape arrive at the multiplexing terminal. These M electrical signals are synchronized, necessary overhead bits are added, and the signals are then interleaved to form a higher rate electrical signal that is usually manufacturer specific.

The electrical multiplexer output signal is clocked out by a local timing signal, either from a local oscillator or a timing signal transported from a remote reference

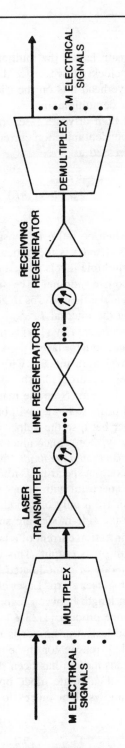

Fig. 1.2 Block diagram of a fiber optic transmission system

clock. Therefore, the output pulse stream leaves the multiplexer with no jitter with respect to the local multiplexer output clock. However, as discussed earlier, the process of synchronizing and inserting overhead bits on the tributary signals results in jitter appearing on the tributaries after demultiplexing.

The high speed electrical signal then drives a laser transmitter that outputs a light signal proportional to a signal dependent injection current. The laser light output signal power in milliwatts can be expressed as

$$l(t) = l_o + \sum_{n=-\infty}^{\infty} a_n m(t - nT) \tag{1.1}$$

where l_o is the optical power output when no pulse is transmitted (e.g. $a_n = 0$), a_n are binary information symbols $a_n \, \varepsilon\{0,1\}$, $m(t)$ is the optical pulse power; and T is the pulse repetition period. Although multiple levels of a_n are possible in fiber optic systems, a_n is usually binary and, of course, non-negative. The output optical pulse of the laser is usually Gaussian shaped and either returns to zero (*RZ coding*) or does not return to zero (*NRZ coding*) during the interval T.

The optical pulse stream with power given by (1.1) is transported on an optical carrier that is not monochromatic. Semiconductor lasers used in most of today's fiber optic systems emit light in several wavelength modes, with the output power generally fluctuating between these modes. Figure 1.3a shows a typical laser output spectrum with center wavelength, $\lambda = 1.3 \, \mu$m. Note the many side-modes. A pulse made up of optical power from several of these modes will broaden in time duration as it propagates through an optical fiber because the velocity of light in the optical fiber is wavelength-dependent. Another type of semiconductor laser that is becoming widely available for system use is the *distributed feedback* (DFB) laser. A DFB laser has suppressed side-modes such that the output spectrum is referred to as single mode. Figure 1.3b shows a typical DFB laser spectrum with center wavelength at 1.5 μm. Light pulses consisting of optical power from only one mode will limit the amount of pulse broadening. However, though the side-modes are suppressed, a DFB laser is not monochromatic. Modulating the injection current of a laser produces variations in both the intensity and the wavelength of its output. This wavelength variation is called *laser chirp*. The single central mode of a modulated DFB laser has spectral width usually less than 0.1 nm caused by laser chirp. Laser chirp will result in pulse broadening at the receiver and the span length between regenerators will be limited by dispersion at a certain bit rate distance product [1.23].

The optical pulse stream from the laser transmitter is coupled into an optical fiber that is usually single mode. A key reason for the emergence of fiber optic transport as the preferred method of transmission has been the rapid improvements in the fabrication and performance of silica single mode optical fiber [1.25]. The low signal attenuation in silica fiber allows light pulses to be transmitted a long

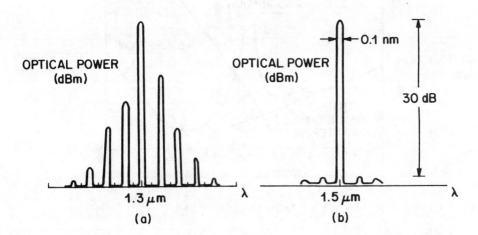

Fig. 1.3 Comparison of a multi-longitudinal mode laser spectrum with a single-longitudinal mode laser spectrum

distance without the need for regeneration. Figure 1.4 shows the loss of a well designed silica single mode optical fiber versus wavelength (λ). Silica fiber has three loss mechanisms: Rayleigh scattering proportional to $1/\lambda^4$ caused by refractive index variations in the non-homogeneous glass, a loss peak near 1.4 μm caused by OH⁻ ion absorption, and a long-wavelength absorption edge beginning near 1.6 μm associated with molecular vibrations in silica. These three intrinsic loss mechanisms combine to result in optical fiber with minimum loss in the 1.5 μm wavelength region. System operation in the 1.5 μm wavelength region will result in the longest span between regenerators.

Most existing fiber optic systems operate in the 1.3 μm wavelength region at the expense of increased loss in order to take advantage of minimum chromatic dispersion. Chromatic dispersion causes a light pulse to broaden in time and arises in a single mode fiber because the velocity of light in a fiber is wavelength-dependent. Since the light pulse is not made up of light of an ideal single wavelength, that is, with a spectrum that can be described mathematically by a delta function; some pulse broadening will occur because of the propagation velocities of the different spectral components. In single mode fibers, the variation in propagation delay is minimal the 1.3 μm wavelength region. In this region, pulse broadening is therefore minimal and lasers with a broad spectrum of output light can be used (Figure 1.3a). However, in

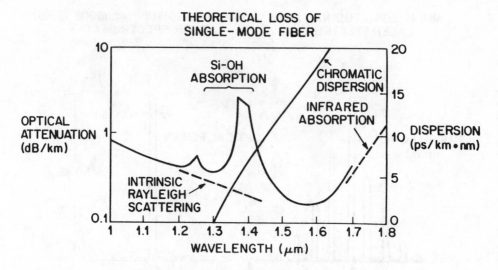

Fig. 1.4 Loss and dispersion of a well designed single mode fiber

the 1.5 μm region, chromatic dispersion is so large that lasers with a narrow spectrum of light output are required to avoid dispersion-related transmission penalties. Figure 1.4 also shows the dispersion coefficient for single mode fiber. Note that at 1.31 μm the dispersion is near zero, while at 1.5 μm the dispersion is greater than 15 picoseconds per kilometer for a light pulse having spectral components spread over one nanometer.

After transmission through the fiber, the optical signal is received by a regenerator and the light signal converted back to an electrical signal using either a PIN or *avalanche photodiode* (APD). In an APD, every primary electron resulting from a photon interaction produces a random number G_{APD} of secondary electrons, hence the name avalanche. A PIN diode receiver, on the other hand, does not produce secondary electrons, therefore $G_{APD} = 1$ and generally is not as sensitive as an APD receiver. Figure 1.5 shows the range of receiver sensitivities versus bit rate for PIN and ADP receivers. The detected photocurrent from the receiver can be expressed as

$$i(t) = q \frac{\lambda e}{hc} \bar{G}_{APD} [l(t) * f(t)] + \eta(t) \qquad (1.2)$$

where q is the quantum efficiency of the device in ampere/watt, $h = 6.62 \times 10^{-34}$ watts (seconds)^2 is Planck's constant, $e = 1.6 \times 10^{-19}$ ampere sec is the electron charge, $c = 3 \times 10^8$ m/sec is the velocity of light in a vacuum and $\bar{G}_{APD}$ is the average gain of the APD. $f(t)$ is the transfer function of the fiber span that is convolved with the light pulse stream $l(t)$. $\eta(t)$ is the receiver noise.

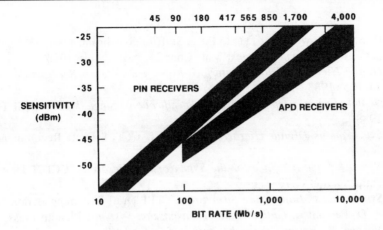

Fig. 1.5 Comparison of PIN and APD receiver sensitivity *versus* bit rate

In a fiber optic transmission system, the receiver noise $\eta(t)$ is composed of two components each with Gaussian distributions: photodiode shot noise and receiver thermal noise. In this book, we will assume that the thermal noise dominates and shot noise is negligible. Once the electrical signal is amplified and reshaped, a timing signal will be extracted from it and a regenerated signal will be produced. This regenerated electrical signal can either go to another laser transmitter that starts the process all over again or to a demultiplexing terminal at the end of the system. At the demultiplexing terminal the higher rate electrical signal is disinterleaved, overhead is removed, and the M electrical signals are recovered.

We will now begin our detailed study of jitter in digital transmission systems by studying the jitter introduced by line regenerators.

REFERENCES

[1.1] A. Papoulis, *Probability, Random Variables and Stochastic Processes*, McGraw Hill, 1965.

[1.2] W. A. Gardner, *Introduction to Random Processes*, MacMillan Publishing Company, 1986.

[1.3] W. R. Bennett, "Statistics of Regenerative Digital Transmission," *Bell System Technical Journal*, Vol. 37, November 1958.

[1.4] E. O. Sunde, "Self-Timing Regenerative Repeaters," *Bell System Technical Journal*, Vol. 36, July 1957.

[1.5] V. I. Johannes and R. H. McCullough, "Multiplexing of Asynchronous Digital Signals Using Pulse Stuffing with Added Bit Signaling," *IEEE Transactions on Communications*, Vol. Com-14, No. 5, October 1966.

[1.6] D. J. Duttweiler, "Waiting Time Jitter," *Bell System Technical Journal*, Vol. 51, No. 1, January 1972.

[1.7]   P. E. K. Chow, "Jitter Due to Pulse Stuffing Synchronization," *IEEE Transactions on Communications*, Vol. Com-21, No. 7, July 1973.

[1.8]   *Digital Hierarchy Optical Interface Rates and Formats Specification*, ANSI T1.105, 1988.

[1.9]   *Digital Hierarchy Optical Interface Specifications: Single-Mode*, ANSI T1.106, 1988.

[1.10] *Synchronous Digital Hierarchy Bit Rates*, CCITT Draft Recommendation G.707.

[1.11] *Network Node Interface for the Synchronous Hierarchy*, CCITT Draft Recommendation G.708.

[1.12] *Synchronous Multiplexing Structure*, CCITT Draft Recommendation G.709.

[1.13] S. D. Personick, *Optical Fiber Transmission Systems*, Plenum Press, 1981.

[1.14] Michael K. Barnoski, *Fundamentals of Optical Fiber Communications*, Academic Press, 1981.

[1.15] P. Halley, *Fiber Optic Systems*, John Wiley & Sons, 1985.

[1.16] D. J. Morris, *Pulse Code Formats for Fiber Optical Data Communications*, Marcel Dekker, Inc., 1983.

[1.17] J. M. Senior, *Optical Fiber Communications*, Prentice-Hall, 1985.

[1.18] J. Gowar, *Optical Communication Systems*, Prentice-Hall, 1984.

[1.19] S. D. Personick, *Fiber Optics*, Plenum Press, 1986.

[1.20] L. B. Jeunhomme, *Single Mode Fiber Optics*, Marcel Dekker, Inc., 1983.

[1.21] S. Geckeler, *Optical Fiber Transmission Systems*, Artech House, 1987.

[1.22] Trondle and Soder, *Optimization of Digital Transmission Systems*, Artech House, 1987.

[1.23] S. E. Miller and I. P. Kaminow, *Optical Telecommunications II,* Academic Press, 1988.

[1.24] P. K. Runge and P. R. Trischitta, *Undersea Lightwave Communications*, IEEE Press, 1986.

[1.25] S. R. Nagel, "Optical Fiber—the Expanding Medium," *IEEE Communications Magazine*, Vol. 25, No. 4, April 1987.

Chapter 2

Jitter Introduced by Line Regenerators

Digital transmission systems, and in particular fiber optic transmission systems, use line regenerators to transport information over longer distances than are possible through a single laser-fiber-receiver span. The line regenerator receives the weak and dispersed incoming optical signal, and attempts to regenerate the original signal as closely as possible for retransmission by its laser transmitter. To regenerate a new signal, a timing signal must be extracted from the received signal. However, the timing extraction process is imperfect, and therefore a timing signal that is inherently jittered will be extracted. The regenerator uses the extracted imperfect timing signal to retransmit a new jittered signal. When regenerators are cascaded to form a digital transmission system, an accumulation of jitter results.

In this chapter, we will study in detail the jitter introduced by line regenerators. We begin by describing the signal processing done inside a regenerator that takes into account receiver noise and jitter from previous regenerators. We then explore the statistics of the received signal to form a mathematical foundation on which to build an understanding of timing extraction. Timing extraction is then discussed in detail, and a jitter model of the regenerator is derived. We examine timing extraction circuit design to understand how to control the jitter that is introduced and transferred by the regenerator. Finally a linear, shift-invariant jitter model of the regenerator is shown to be experimentally valid for fiber optic regenerators. This chapter provides the analytical and experimental foundations for Chapter 3 on the accumulation of jitter for a chain of cascaded regenerators.

2.1 FIBER OPTIC REGENERATOR DESCRIPTION

In Figure 2.1 we show the block diagram of a self-timed fiber optic regenerator [2.1]. An optical receiver, containing a PIN or an APD, converts the optical signal from the incoming single-mode optical fiber into an electrical signal. This electrical signal is then passed through an equalizer and an *automatic gain control* (AGC) amplifier

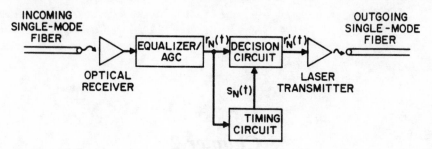

Fig. 2.1 Block diagram of a self-timed fiber optic regenerator

that outputs a signal $r_N(t)$. The signal $r_N(t)$, received by the Nth regenerator in a long chain of cascaded regenerators, is corrupted by receiver noise and jitter accumulated from the previous $N - 1$ regenerators. We express $r_N(t)$ as

$$r_N(t) = \sum_{n=-\infty}^{\infty} a_n g_N(t - nT - e_{N-1}[nT]) + \eta_N(t) \tag{2.1}$$

where $a_n \in \{0,1\}$, $g_N(t)$ is the amplified and equalized received pulse, $\eta_N(t)$ is additive receiver noise, and $e_{N-1}[nT]$ is the time deviation from time nT of the nth pulse's zero crossing (See Figure 1.1). The $e_{N-1}[nT]$ jitter sequence is the accumulated jitter from the previous $N - 1$ regenerators.

The received signal is sampled by the decision circuit and a regenerated output signal is produced based on decisions made about $r_N(t)$ at the sampling instances. The timing signal used for sampling comes from the timing extraction circuit, which extracts a sine wave of a frequency $1/T$ from $r_N(t)$. Since timing extraction is imperfect, the timing signal is given by

$$s_N(t) = |A_N| \sin \frac{2\pi}{T} (t - e_N(t)) \tag{2.2}$$

where $|A_N|$ is a constant amplitude, and $e_N(t)$ is the jitter on the timing signal. Note that $e_N(t)$ is a continuous time phase modulation of $s_N(t)$; $e_N(t)$ is caused by imperfections of the timing extraction process, the input jitter $e_{N-1}[nT]$, and the receiver noise $\eta_N(t)$.

When $s_N(t)$ rises to a voltage V, $r_N(t)$ is sampled and the decision circuit creates a regenerated output signal, that in turn modulates a semiconductor laser transmitter [2.2]. The regenerated output signal is given by

$$r_N'(t) = \sum_{n=-\infty}^{\infty} a_n' f_N(t - nT - e_N[nT] + \tau + t_o) \tag{2.3}$$

where a'_n is the regenerated bit sequence ($a'_n = a_n$ if a correct decision was made), $f_N(t)$ is the output pulse, t_o and τ are static phase shifts to be discussed in Chapter 4, and $e_N[nT]$ is the output jitter of the *Nth* regenerator, a sampled $e_N(t)$. Therefore, the timing signal jitter is impressed onto the output data signal.

Equations (2.1) through (2.3) mathematically describe the signal processing done inside a line regenerator. Note that jitter is present on both the input and the output signals. To describe how jitter is introduced by the line regenerators, and how input jitter passes through the regenerator, we must study the timing extraction process in detail. However, before analyzing the timing extraction process, it is important that we explore the structure of the received signal given by (2.1). This will aid in our understanding of the timing extraction process, to be described in later sections of this chapter.

2.2 THE CYCLOSTATIONARY RECEIVED DATA SIGNAL

By first considering the statistics of the received data signal, $r_N(t)$, we can form a foundation on which to build an understanding of timing extraction. This was first done by Bennett in [2.3]. In this section, we will study a noiseless, jitterless, received data signal. However, using a noisy, jittered received signal would not change our results.

We calculate the mean of $r_N(t)$ as

$$E\{r_N(t)\} = E\{a_n\} \sum_{n=-\infty}^{\infty} g_N(t - nT) \tag{2.4}$$

and its autocorrelation as

$$R_{rr}(t, t + \tau) = E\{r_N(t)r_N^*(t + \tau)\}$$

$$= E\left\{ \sum_{n=-\infty}^{\infty} a_n g_N(t - nT) \sum_{l=-\infty}^{\infty} a_l g_N(t - lT + \tau) \right\} \tag{2.5}$$

$$= E\left\{ \sum_{n=-\infty}^{\infty} \sum_{m=-\infty}^{\infty} a_n a_{n+m} g_N(t - nT) g_N(t - nT - mT + \tau) \right\} \tag{2.6}$$

$$= \sum_{n=-\infty}^{\infty} \sum_{m=-\infty}^{\infty} E\{a_n a_{n+m}\} g_N(t - nT) g_N(t - nT - mT + \tau) \tag{2.7}$$

Note that both the mean and autocorrelation of $r_N(t)$ are periodic with period T. A stochastic process with a periodic mean and autocorrelation is referred to as *cyclostationary*. Because $r_N(t)$ is cyclostationary with period T, a timing signal with period T can be extracted from it [2.4]. Before discussing how timing extraction is done, let us calculate the average power spectrum of $r_N(t)$.

To find the average power spectrum of a cyclostationary stochastic process, the time dependence of $R_{rr}(t,t + \tau)$ can be eliminated by averaging over the period T [2.7]. This results in an average autocorrelation of

$$\bar{R}_{rr}(\tau) = \frac{1}{T} \int_{-T/2}^{T/2} R_{rr}(t,t + \tau)dt \tag{2.8}$$

$$= \sum_{n=-\infty}^{\infty} \sum_{m=-\infty}^{\infty} E\{a_n a_{n+m}\} \frac{1}{T} \int_{-T/2}^{T/2}$$
$$\cdot g_N(t - nT)g_N(t - nT - mT + \tau)dt \tag{2.9}$$

$$= \sum_{m=-\infty}^{\infty} \sum_{n=-\infty}^{\infty} E\{a_n a_{n+m}\} \frac{1}{T} \int_{-T/2-nT}^{T/2-nT}$$
$$\cdot g_N(t)g_N(t - mT + \tau)dt \tag{2.10}$$

Since a_n is wide sense stationary, $E\{a_n a_{n+m}\}$ does not depend on n; therefore

$$\bar{R}_{rr}(\tau) = \sum_{m=-\infty}^{\infty} E\{a_n a_{n+m}\} \frac{1}{T} \int_{-\infty}^{\infty} g_N(t)g_N(t - mT + \tau)dt \tag{2.11}$$

Examining (2.11) we note that the integral is the autocorrelation of $g_N(t)$, and the first term is the autocorrelation of a_n. By defining $R_{gg}(\tau)$ as the autocorrelation of $g_N(t)$, and $R_{aa}[m]$ as the autocorrelation of a_n, (2.11) becomes

$$\bar{R}_{rr}(\tau) = \frac{1}{T} \sum_{m=-\infty}^{\infty} R_{aa}[m]R_{gg}(\tau - mT)$$

$$= \frac{1}{T} R_{gg}(\tau) * \sum_{m=-\infty}^{\infty} R_{aa}[m]\delta(\tau - mT) \tag{2.12}$$

Fourier transforming (2.12) yields the average power spectrum of $r_N(t)$ as

$$\bar{S}_{rr}(f) = \frac{1}{T} |G_N(f)|^2 S_{aa}(f) \tag{2.13}$$

where $G_N(f)$ is the Fourier transform of $g_N(t)$, and $S_{aa}(f)$ is the power spectrum of the message sequence $\{a_n\}$ defined as the discrete Fourier transform of $R_{aa}[m]$, i.e.

$$S_{aa}(f) = \sum_{m=-\infty}^{\infty} R_{aa}[m]e^{-j2\pi fmT} \tag{2.14}$$

Examining (2.13) shows that the average power spectrum of $r_N(t)$ is dependent on the equalized pulse shape and the second order statistics of the message sequence. The effect of message sequence statistics will be examined first for two cases.

Case I: If the message sequence, $\{a_n\} \in \{0,1\}$ is independent, then $R_{aa}[m] = 1/4\, \delta[m] + 1/4$ and

$$S_{aa}(f) = \frac{1}{4} + \frac{1}{4} \sum_{m=-\infty}^{\infty} e^{-j2\pi fmT}$$

$$= \frac{1}{4} + \frac{1}{4T} \sum_{m=-\infty}^{\infty} \delta\left(f - \frac{m}{T}\right) \qquad (2.15)$$

Inserting (2.15) into (2.13) yields an average power spectrum of

$$\bar{S}_{rr}(f) = \frac{1}{4T}|G_N(f)|^2 + \frac{1}{4T^2} \sum_{m=-\infty}^{\infty} \left|G_N\left(\frac{m}{T}\right)\right|^2 \delta(f - mT) \qquad (2.16)$$

Therefore, for an independent message sequence, $\bar{S}_{rr}(f)$ has a continuous spectra of just the pulse shape magnitude squared plus discrete components at $f = m/T$. Depending on $|G_N(1/T)|$, we could use the discrete component at $f = 1/T$ as a timing signal; however, as we will see later, $|G_N(1/T)|$ is usually small to conserve bandwidth in high bit rate transmission systems.

Case II: If the message sequence has some periodicity, such as a bit sequence produced by a shift register pattern generator [2.6], $R_{aa}[m]$ will be periodic with period v bits. For periodic shift register patterns that output zeroes and ones, the autocorrelation is given by

$$R_{aa}[m] = \frac{1}{4}\left(1 + \frac{1}{v}\right)\delta[m] + \frac{1}{4}\left(1 - \frac{1}{v}\right)$$

for $m = 0$ to v and

$$R_{aa}[m + kv] = R_{aa}[m] \qquad (2.17)$$

for all integer k. Inserting (2.17) into (2.14) results in

$$S_{aa}(f) = \frac{1}{4T}\left(1 + \frac{1}{v}\right)\sum_{k=-\infty}^{\infty}\delta\left(f - \frac{k}{vT}\right) + \frac{1}{4T}\left(1 - \frac{1}{v}\right)\sum_{m=-\infty}^{\infty}\delta\left(f - \frac{m}{T}\right) \qquad (2.18)$$

Substituting (2.18) into (2.13) yields an average power spectrum of $r_N(t)$ of

$$\bar{S}_{rr}(f) = \frac{1}{4T}\left(1 + \frac{1}{\nu}\right)\sum_{k=-\infty}^{\infty}\left|G_N\left(\frac{k}{\nu T}\right)\right|^2\delta\left(f - \frac{k}{\nu T}\right)$$

$$+ \frac{1}{4T^2}\left(1 - \frac{1}{\nu}\right)\sum_{m=-\infty}^{\infty}\left|G_N\left(\frac{m}{T}\right)\right|^2\delta(f - mT) \qquad (2.19)$$

Therefore for a periodic shift register pattern, $\bar{S}_{rr}(f)$ has a discrete spectra with impulses at $f_k = k/\nu T$. The magnitude of the impulses are shaped by the pulse shape magnitude squared.

Comparing (2.19) and (2.16), we see that in the limit as $\nu \to \infty$ (2.19) is equivalent to (2.16). In other words, the spectrum of a periodic message sequence consists of impulses spaced $1/\nu T$ Hz apart and as $\nu \to \infty$ this portion of the power spectrum becomes a continuum. These two cases illustrated the effect of message statistics on the received signal's power spectrum.

Usually the message sequence is coded or bits are added to the message sequence for signaling, framing, *et cetera*. The coding of the message sequence and the adding of a periodic framing pattern changes the received signal's power spectrum. We have discussed two simple cases of how the message sequence effects $\bar{S}_{rr}(f)$. These were chosen because of ease of analysis and because we often use a periodic input sequence to approximate an independent message sequence. With ν large, say $\nu = 2^{23} - 1$ bits, a shift register pattern adequately approximates an independent message sequence.

Next we will consider the pulse shape. To limit the amount of intersymbol interference, and therefore decrease the probability of making a bit error, the pulse shape $g_N(t)$ is usually chosen from the Nyquist family of pulse shapes [2.7]. Choosing

$$g_N(t) = \frac{\sin \pi t/T}{\pi t/T}\frac{\cos \dfrac{\pi t}{T}}{1 - 4t^2/T^2}$$

with transform:

$$G_N(f) = \frac{T}{2}\left[1 - \sin \pi T\left(f - \frac{1}{2T}\right)\right]$$

for $0 < f < 1/T$, (2.13) is graphed in Figure 2.2 for an independent message signal (2.16) and for a periodic message signal (2.19). It is important to note that at the bit repetition frequency, $1/T$, there is no power in either case. Therefore, when a

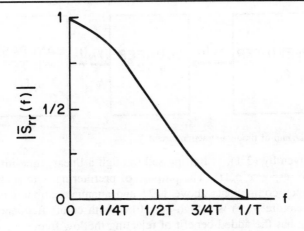

INDEPENDENT MESSAGE SEQUENCE

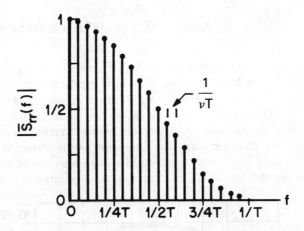

Fig. 2.2 Average power spectrum of received signal with (a) independent message sequence and (b) periodic message sequence

raised cosine pulse shape is used, a timing signal of frequency $1/T$ must be extracted from a signal without average power at $f = 1/T$. In the next section we will analyze the time extraction process in detail and show that we can extract a timing signal from $r_N(t)$ regardless of power at $f = 1/T$, because $r_N(t)$ is cyclostationary with period T.

2.3 TIMING EXTRACTION

Figure 2.3 shows the block diagram of the timing extraction circuit. The received

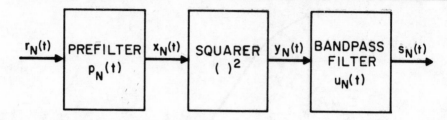

Fig. 2.3 Block diagram of timing extraction circuit

signal $r_N(t)$, given by (2.1), is first passed through a linear, time-invariant prefilter having impulse response $p_N(t)$. The purpose of prefiltering is to shape $r_N(t)$ to emphasize frequency components above $1/2T$ and attenuate components below $1/2T$ [2.8, 2.9]. Therefore $p_N(t)$ is a high-pass filter with cutoff frequency near $f = 1/2T$. Prefiltering has the added benefit of rejecting the low frequency components of $\eta_N(t)$ that enter the timing circuit. The output of the prefilter is given by

$$x_N(t) = \sum_{n=-\infty}^{\infty} a_n q_N(t - nT - e_{N-1}[nT]) + \eta_N(t) * p_N(t) \qquad (2.20)$$

where $q_N(t) = p_N(t) * g_N(t)$. Figure 2.4 shows a measured prefilter output spectrum. Note that $x_N(t)$ does not have a strong discrete component at $f = 1/T$ or $f = 0$. However, since $x_N(t)$ is cyclostationary with period T, we can extract a discrete component at $f = 1/T$ by passing $x_N(t)$ through a non-linear circuit. By assuming a square-law nonlinearity, the analysis is tractable, whereas methods considering other nonlinear circuits rely more on computer simulation of the extracted timing signal. Although not discussed here, other nonlinearities have been considered [2.10–2.13] and have shown that the square law is a desirable nonlinearity for timing extraction.

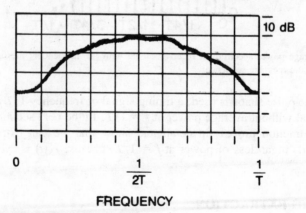

Fig. 2.4 Measure prefilter output spectrum

So, passing $x_N(t)$ through a squarer results in

$$y_N(t) = x_N^2(t) = \sum_{n=-\infty}^{\infty} \sum_{m=-\infty}^{\infty} a_n a_{n+m} q_N(t - nT - e_{N-1}[nT])$$

$$\cdot q_N(t - nT - mT - e_{N-1}[(n + m)T])$$

$$+ 2[\eta_N(t) * p_N(t)] \sum_{n=-\infty}^{\infty} a_n q_N(t - nT - e_{N-1}[nT]) \qquad (2.21)$$

$$+ [\eta_N(t) * p_N(t)]^2$$

Examining (2.21), we see that the squarer output consists of three terms with the last two describing how receiver noise passes through the timing circuit. Since these terms do not contain timing information, we will concentrate on the first term. To show that the first term contains a discrete component at $f = 1/T$, we will set the input jitter to zero and find the power spectrum of

$$y_N(t) = \sum_{n=-\infty}^{\infty} \sum_{m=-\infty}^{\infty} a_n a_{n+m} q_N(t - nT) q_N(t - nT - mT) \qquad (2.22)$$

The autocorrelation of $y_N(t)$ is

$$R_{yy}(t + \tau, t) = E\{y_N(t) y_N^*(t + \tau)\}$$

$$= E\Bigg\{ \sum_{n=-\infty}^{\infty} \sum_{m=-\infty}^{\infty} a_n a_{n+m} q_N(t - nT) q_N(t - nT - mT)$$

$$\cdot \sum_{l=-\infty}^{\infty} \sum_{k=-\infty}^{\infty} a_l a_{l+k} q_N(t - lT + \tau) q_N(t - lT - kT + \tau) \Bigg\}$$

$$= \sum_{n} \sum_{m} \sum_{l} \sum_{k} E\{a_n a_{n+m} a_{n+l} a_{n+l+k}\}$$

$$\cdot q_N(t - nT) q_N(t - nT - mT) q_N(t + \tau - nT - lT)$$

$$\cdot q_N(t + \tau - nT - lT - kT) \qquad (2.23)$$

A zero-mean message sequence simplifies our analysis by removing the discrete component at $f = 1/T$ that would be present [2.8]. Hence, assuming an independent message sequence with zero mean, then

$$E\{a_n a_{n+m} a_{n+l} a_{n+l+k}\} = 1, \text{ if } \quad m=l=k=0$$
$$m=k=0, l\neq 0$$
$$m=l\neq 0, l=-k$$
$$m=k\neq 0, l=0$$
$$= 0, \text{ otherwise.} \tag{2.24}$$

Using (2.24) to sort out the non-zero terms (2.23) simplifies to

$$R_{yy}(t + \tau, t) = \sum_{n=-\infty}^{\infty} q_N^2(t - nT)q_N^2(t - nT + \tau)$$

$$+ \sum_{n=-\infty}^{\infty} \sum_{l=-\infty}^{\infty} q_N^2(t - nT)q_N^2(t - lT - nT + \tau)$$

$$+ 2 \sum_{n=-\infty}^{\infty} \sum_{l=-\infty}^{\infty} q_N(t - nT)q_N(t - nT - lT)$$

$$\cdot q_N(t - nT + \tau)q_N(t - nT - lT + \tau) \tag{2.25}$$

Since $y_N(t)$ is cyclostationary with period T, we can remove the time dependence of $R_{yy}(t + \tau, t)$ by integrating over a period T. This results in the average correlation function of

$$\bar{R}_{yy}(\tau) = \frac{1}{T} \int_{-T/2}^{T/2} R_{yy}(t + \tau, t)dt$$

$$= \frac{1}{T} \int_{-\infty}^{\infty} q_N^2(t)q_N^2(t + \tau)dt + \frac{1}{T} \sum_{l=-\infty}^{\infty} \int_{-\infty}^{\infty} q_N^2(t)q_N^2(t - lT + \tau)dt$$

$$+ \frac{2}{T} \sum_{l=-\infty}^{\infty} \int_{-\infty}^{\infty} q_N(t)q_N(t - lT)q_N(t + \tau)q_N(t - lT + \tau)dt \tag{2.26}$$

By defining $R_{q^2q^2}(\tau)$ as the autocorrelation of $q_N^2(t)$, $\bar{R}_{yy}(\tau)$ becomes

$$\bar{R}_{yy}(\tau) = \frac{1}{T} R_{q^2q^2}(\tau) + \frac{1}{T} \sum_{l=-\infty}^{\infty} R_{q^2q^2}(\tau - lT)$$

$$+ \frac{2}{T} R_{q^2q^2}(\tau) * \sum_{m=-\infty}^{\infty} \delta(\tau - mT)$$

$$= \frac{1}{T} R_{q^2q^2}(\tau) + \frac{3}{T} R_{q^2q^2}(\tau) * \sum_{l=-\infty}^{\infty} \delta(\tau - lT) \tag{2.27}$$

The Fourier transform of (2.27) yields an average power spectrum of $y_N(t)$ as

$$\bar{S}_{yy}(f) = \frac{1}{T} S_{q^2q^2}(f) + \frac{3}{T^2} S_{q^2q^2}\left(\frac{l}{T}\right) \sum_{l=-\infty}^{\infty} \delta\left(f - \frac{l}{T}\right)$$

(2.28)

Examining (2.28) shows that $\bar{S}_{yy}(f)$ contains a continuous spectra plus discrete components at frequencies $f = l/T$. The discrete components are present because $r_N(t)$ is cyclostationary. In Figure 2.5, we show a measured $\bar{S}_{yy}(f)$ of a fiber optic regenerator. Note the agreement with (2.28), including the strong discrete component at $f = 1/T$.

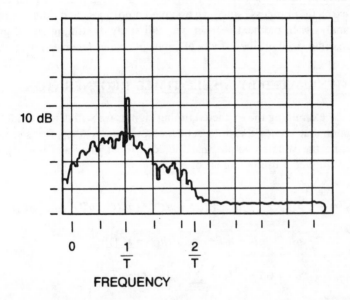

Fig. 2.5 Squarer output spectrum

Passing $y_N(t)$ through a narrow band bandpass filter with impulse response $u_N(t)$ results in an extracted timing signal of

$$s_N(t) = y_N(t) * u_N(t)$$

(2.29)

Since $u_N(t)$ with transform $U_N(f)$ is a narrow band bandpass filter centered near $f = 1/T$, $U(l/T) = 0$ for $l \neq \pm 1$. This results in a timing signal average power spectrum of

$$\bar{S}_{ss}(f) = \frac{1}{T} \left| U_N(f) \right|^2 S_{q^2q^2}(f)$$

$$+ \frac{3}{T^2} \left| U_N\left(\frac{1}{T}\right) \right|^2 S_{q^2q^2}\left(\frac{1}{T}\right) \left[\delta\left(f - \frac{1}{T}\right) + \delta\left(f + \frac{1}{T}\right) \right] \quad (2.30)$$

Equation (2.30) shows that, in addition to discrete components at $f = \pm 1/T$, the timing signal also contains unwanted power around $f = 1/T$, (i.e., jitter). Thus we have shown in this section that the spectrum of the extracted signal contains a discrete component at $f = 1/T$. However, it also contains an unwanted continuous spectrum that results in jitter being on the extracted timing signal. Note that this jitter is caused by the timing extraction process, since we assumed a noiseless and jitterless received signal in this analysis. In the next section, we will study the timing signal given by (2.29) that takes into account the effects of input jitter and receiver noise.

2.4 JITTER GENERATED BY A SELF-TIMED REGENERATOR

In this section, by examining the extracted timing signal, we will derive a jitter model of the regenerator that relates the jitter sequence on $r_N(t)$ to the jitter sequence on $r'_N(t)$. Substituting the squarer output that includes input jitter and noise given by (2.21) into (2.29) gives a timing signal of

$$s_N(t) = \sum_{n=-\infty}^{\infty} \sum_{m=-\infty}^{\infty} a_n a_{n+m} [u_N(t - nT) * [q_N(t - nT - e_{N-1}[nT])]$$

$$\cdot q_N(t - nT - mT - e_{N-1}[(n + m)T])]]$$

$$+ 2[\eta_N(t) * p_N(t)] \sum_{n=-\infty}^{\infty} a_n [u_N(t - nT)$$

$$* q_N(t - nT - e_{N-1}[nT])]$$

$$+ [\eta_N(t) * p_N(t)]^2 * u_N(t) \quad (2.31)$$

Noting that $s_N(t, e_{N-1}[nT], e_{N-1}[(n + m)T])$ is a function of three variables, $s_N(t)$ can be approximated by the first terms of the two dimensional power series expansion of the variables $e_{N-1}[nT]$ and $e_{N-1}[(n + m)T]$, [2.14]. We obtain

$$s_N(t) \approx \sum_{n=-\infty}^{\infty} \sum_{m=-\infty}^{\infty} a_n a_{n+m} [u_N(t - nT) * [q_N(t - nT)q_N(t - nT - mT)]]$$

$$- 2 \sum_{n=-\infty}^{\infty} \sum_{m=-\infty}^{\infty} a_n a_{n+m} e_{N-1}[nT][u_N(t - nT)$$

$$* q'_N(t - nT)q_N(t - nT - mT)]$$

$$+ 2[\eta_N(t) * p_N(t)] \sum_{n=-\infty}^{\infty} a_n[u_N(t - nT) * q_N(t - nT)]$$

$$- 2[\eta_N(t) * p_N(t)] \sum_{n=-\infty}^{\infty} a_n e_{N-1}[nT][u_N(t - nT) * q_N'(t - nT)]$$

$$+ [\eta_N(t) * p_N(t)]^2 * u_N(t) \tag{2.32}$$

where

$$q_N'(t) = \frac{d}{dt} q_N(t)$$

The timing signal given by (2.32) can be shown to be a jittered sine wave of period T. By comparing $s_N(t)$ given by (2.32), with the mean timing signal, an expression for the jitter on $s_N(t)$ can be derived.

The mean of $s_N(t)$ is

$$E\{s_N(t)\} = \sum_{n=-\infty}^{\infty} \sum_{m=-\infty}^{\infty} E\{a_n a_{n+m}\}[u_N(t - nT)$$

$$* q_N(t - nT)q_N(t - nT - mT)]$$

$$- 2 \sum_{n=-\infty}^{\infty} \sum_{m=-\infty}^{\infty} E\{a_n a_{n+m} e_{N-1}[nT]\}[u_N(t - nT)$$

$$* q_N'(t - nT)q_N(t - nT - mT)]$$

$$+ 2[E\{\eta_N(t)\} * p_N(t)] \sum_{n=-\infty}^{\infty} E\{a_n\}$$

$$\cdot [u_N(t - nT) * q_N(t - nT)]$$

$$- 2[E\{\eta_N(t)\} * p_N(t)] \sum_{n=-\infty}^{\infty} E\{a_n e_{N-1}[nT]\}$$

$$\cdot [u_N(t - nT) * q_N'(t - nT)]$$

$$+ E\{[\eta_N(t) * p_N(t)]\}^2 * u_N(t) \tag{2.33}$$

For a zero-mean $\eta_N(t)$, an independent data signal $a_n \in \{0,1\}$ such that

$$\sum_{m=-\infty}^{\infty} E\{a_n a_{n+m}\} = \frac{1}{4} \delta[m] + \frac{1}{4}$$

(case I, section 2.2), and recalling that $u_N(t)$ is bandpass, (2.33) is simplified to

$$E\{s_N(t)\} = \frac{1}{4} \sum_{n=-\infty}^{\infty} u_N(t - nT) * q_N^2(t - nT)$$

$$+ \frac{1}{4} \sum_{n=-\infty}^{\infty} \sum_{m=-\infty}^{\infty} u_N(t - nT) * q_N(t - nT)q_N(t - nT - mT)$$

$$- 2 \sum_{n=-\infty}^{\infty} \sum_{m=-\infty}^{\infty} b_m [u_N(t - nT)$$

$$* q_N'(t - nT)q_N(t - nT - mT)] \tag{2.34}$$

where $b_m = E\{a_n a_{n+m} e_{N-1}[nT]\}$ does not depend on n. Applying the Poisson Summation Formula [2.15] to (2.34) results in

$$E\{s_N(t)\} = \frac{1}{4T} \sum_{l=-\infty}^{\infty} U_N\left(\frac{l}{T}\right) \left[Q_N\left(\frac{l}{T}\right) * Q_N\left(\frac{l}{T}\right) \right] e^{j(2\pi/T)lt}$$

$$+ \frac{1}{4T} \sum_{m=-\infty}^{\infty} \sum_{l=-\infty}^{\infty} U_N\left(\frac{l}{T}\right) \int_{-\infty}^{\infty} Q_N(f)Q\left(\frac{l}{T} - f\right)$$

$$\cdot e^{-j2\pi(l/T-f)mT} df e^{j(2\pi/T)lt} - \frac{2}{T} \sum_{m=-\infty}^{\infty} b_m \sum_{l=-\infty}^{\infty} U_N\left(\frac{l}{T}\right)$$

$$\cdot \int_{-\infty}^{\infty} Q_N(f)Q_N\left(\frac{l}{T} - f\right) e^{-j2\pi(l/T-f)mT} j2\pi f df e^{j(2\pi/T)lt} \tag{2.35}$$

where $U_N(f)$ and $Q_N(f)$ are the Fourier transforms of $u_N(t)$ and $q_N(t)$, respectively. Since $U_N(f)$ is narrow band centered near $f = 1/T$, $U_N (l/T) = 0$ for $l \neq 1, -1$. Applying this to (2.35) results in a sinusoidal mean timing signal of

$$E\{s_N(t)\} = |A_N| \cos\left(\frac{2\pi}{T} t + \arg A_N\right)$$

where

$$A_N = \frac{1}{4T} U_N\left(\frac{1}{T}\right) \left[Q_N\left(\frac{1}{T}\right) * Q_N\left(\frac{1}{T}\right) \right]$$

$$+ \frac{1}{4T} \sum_{m=-\infty}^{\infty} U_N\left(\frac{1}{T}\right) \int_{-\infty}^{\infty} Q_N(f)Q_N\left(\frac{1}{T} - f\right) e^{-j2\pi(1/T-f)mT} df$$

$$- \frac{2}{T} \sum_{m=-\infty}^{\infty} b_m U_N\left(\frac{1}{T}\right) \int_{-\infty}^{\infty} Q_N(f)Q_N\left(\frac{1}{T} - f\right) j2\pi f e^{-j2\pi(1/T-f)mT} df \tag{2.36}$$

Examining (2.36) we see that $E\{s_N(t)\}$ the mean timing signal has rising zero crossing at times

$$t_k = \frac{3T}{4} - \frac{T}{2\pi} \arg A_N + kT$$

Shifting the time axis such that the rising zero crossings of $E\{s_N(t_k')\}$ are at multiples of kT, we graph in Figure 2.6 $E\{s_N(t_k')\}$ and $s_N(t')$ at $t' = kT$. We define the output jitter of the regenerator, $e_N[kT]$, as the difference between $E\{s_N(kT)\}$ and $s_N(kT)$. If $e_N[kT]$ is small, the following relationship results [2.16]

$$e_N[kT] \approx \frac{s_N[kT] - E\{s_N[kT]\}}{\dfrac{d}{dt'} E\{s_N(t')\}|_{t'=kT}} \tag{2.37}$$

Substituting (2.32) and (2.34) into (2.37) yields an output jitter of

$$
\begin{aligned}
e_N[kT] \approx &\frac{T}{2\pi|A_N|} \sum_{n=-\infty}^{\infty} \sum_{m=-\infty}^{\infty} \left[a_n a_{n+m} - \left(\frac{1}{4}\delta[m] + \frac{1}{4} \right) \right] \\
&\cdot [u_N(kT - nT) * q_N(kT - nT)q_N(kT - nT - mT)] \\
&- \frac{T}{\pi|A_N|} \sum_{n=-\infty}^{\infty} \sum_{m=-\infty}^{\infty} [a_n a_{n+m} e_{N-1}[nT] - b_m] \\
&\cdot [u_N(kT - nT) * q_N'(kT - nT)q_N(kT - nT - mT)] \\
&+ \frac{T}{\pi|A_N|} [\eta_N[kT] * p_N[kT]] \sum_{n=-\infty}^{\infty} \\
&\cdot a_n[u_N(kT - nT) * q(kT - nT)] \\
&- \frac{T}{\pi|A_N|} [\eta_N[kT] * p_N[kT]] \sum_{n=-\infty}^{\infty} a_n e_{N-1}[nT] \\
&\cdot [u_N(kT - nT) * q_N'(kT - nT)] \\
&+ \frac{T}{2\pi|A_N|} [[\eta_N[kT] * p_N[kT]]^2 \\
&- E\{[\eta_N[kT] * p_N[kT]]^2\}] * u_N[kT]
\end{aligned} \tag{2.38}
$$

Examining (2.38) we can express $e_N[kT]$ as several terms each containing a discrete convolution ⊞ i.e.

$$e_N[kT] = \frac{2T}{\pi} \sum_{m=-\infty}^{\infty} \left[\left[a_k a_{k+m} - \left(\frac{1}{4} \delta[m] + \frac{1}{4} \right) \right] \circledast \frac{1}{4|A_N|} [u_N[kT] \right.$$

$$\left. * q_N[kT]q_N[kT - mT]] \right]$$

$$- \frac{4T}{\pi} \sum_{m=-\infty}^{\infty} \left[(a_k a_{k+m} e_{N-1}[kT] - b_m) \circledast \frac{1}{4|A_N|} [u_N[kT] \right.$$

$$\left. * q_N'[kT]q_N[kT - mT]] \right]$$

$$+ \frac{4T}{\pi} [\eta_N(kT) * p_N(kT)]a_k \circledast \frac{1}{4|A_N|} [u_N[kT] * q_N[kT]]$$

$$- \frac{4T}{\pi} [\eta_N(kT) * p_N(kT)]a_k e_{N-1}[kT]$$

$$\circledast \frac{1}{4|A_N|} [u_N[kT] * q_N'[kT]]$$

$$+ \frac{2T}{\pi} [[\eta_N(kT) * p_N(kT)]^2 - E\{[\eta_N(kT)$$

$$* p_N(kT)]^2\}] \circledast \frac{1}{4|A_N|} u_N[kT] \qquad (2.39)$$

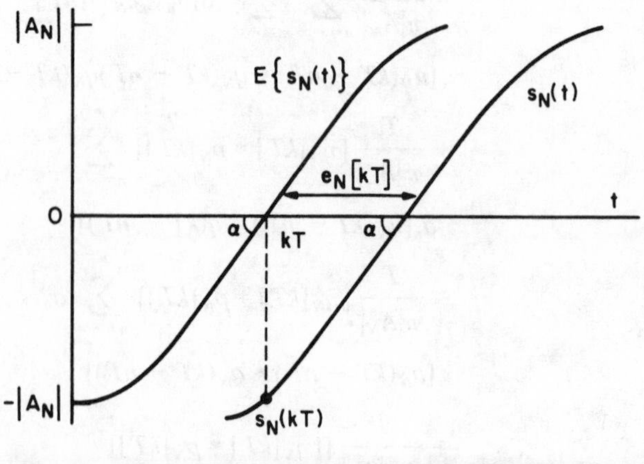

Fig. 2.6 Defining jitter geometrically by examining $E\{s_N(t)\}$ and $s_N(t)$ near $t = kT$

In the next section on the jitter transfer function we show that the right side of each discrete convolution in (2.39) can be represented by a linear jitter filter with impulse response $h_N[kT]$. We define the jitter generated by the regenerator that is pattern-dependent but noise-independent as

$$e_{iN}^P[kT] \underset{=}{\Delta} \frac{2T}{\pi} \sum_{m=-\infty}^{\infty} \left(a_k a_{k+m} - \left(\frac{1}{4} \delta[m] + \frac{1}{4} \right) \right) \qquad (2.40\,\text{a})$$

the jitter generated that is pattern-dependent and noise-dependent as

$$e_{iN}^{\eta P}[kT] \underset{=}{\Delta} \frac{4T}{\pi} [\eta_N(kT) * p_N(kT)](a_k - a_k e_{N-1}[kT]) \qquad (2.40\,\text{b})$$

and the jitter that is noise-dependent but pattern-independent as

$$e_{iN}^{\eta}[kT] \underset{=}{\Delta} \frac{2T}{\pi} [\eta_N(kT) * p_N(kT)]^2 - E\{[\eta_N(kT) * p_N(kT)]^2\} \qquad (2.40\,\text{c})$$

The three types of jitter internally generated by the regenerator given by (2.40 a, b, and c) can be shown to be zero mean and white stochastic processes, under the assumption of an independent message sequence $a_n \in \{0,1\}$ and zero mean white noise process $\eta_N(t)$.

The slowly varying input jitter $e_{N-1}[kT]$ is present only in the second term of (2.39), and is modulated by the broadband $\sum_{m=-\infty}^{\infty} a_k a_{k+m}$ message sequence. Only the $m = 0$ term will pass through the bandpass filter, therefore

$$a_k^2 e_{N-1}[kT] + E\{a_k^2 e_N[kT]\} = e_{N-1}[kT]$$
$$+ [[a_k^2 - 1]e_{N-1}[kT] - E\{[a_k^2 - 1]e_{N-1}[kT]\}] \qquad (2.40\,\text{d})$$

where the second term is broadband, zero mean and can be ignored when an independent message signal is received.

Using the definitions given by (2.40), together with $h_N[kT]$ in (2.39) yields a jitter model for a regenerator of

$$e_N[kT] = \left[e_{iN}^P[kT] + e_{iN}^{\eta P}[kT] + e_{iN}^{\eta}[kT] + e_{N-1}[kT] \right] \circledast h_N[kT] \qquad (2.41)$$

Therefore, we have shown that the output jitter of the regenerator is a filtered sum of the input jitter plus the jitter generated internally by the regenerator. Figure 2.7 illustrates this jitter model of a regenerator [2.17–2.20]. In the following sections, we will show experimentally that this linear, shift-invariant jitter model is valid for fiber optic regenerators. But before validating the jitter model of the complete regenerator, let us look in detail at the jitter filter given by $h_N[kT]$.

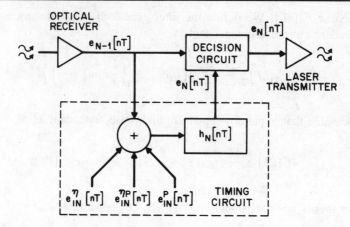

Fig. 2.7 Jitter model of a fiber optic regenerator

2.5 THE JITTER TRANSFER FUNCTION

In deriving the jitter model of the regenerator, we said that the jitter generated by the regenerator and the incoming jitter to the regenerator are filtered by a linear, shift-invariant jitter filter described by $h_N[kT]$. In this section we will explore this jitter filter in detail. Examining the second terms of the discrete convolutions in (2.39) we will define the jitter filter impulse response as

$$h_N^m[kT] \triangleq \frac{1}{4|A_N|} [u_N[kT] * q_N'[kT]q_N[kT - mT]] \qquad (2.42)$$

We also define the jitter transfer function as the discrete Fourier transform of (2.42), i.e.,

$$H_N^m(f) = \sum_{k=-\infty}^{\infty} h_N^m[kT]e^{-j2\pi fkT} \qquad (2.43)$$

Using the Poisson Sum Formula and (2.42) results in

$$H_N^m(f) = \frac{1}{4|A_N|T} \sum_{l=-\infty}^{\infty} U_N\left(f + \frac{l}{T}\right) \int_{-\infty}^{\infty} q_N'(t)q_N(t - mT)e^{j2\pi(f+l/T)t}dt \qquad (2.44)$$

where $U_N(f)$ is the Fourier transform of $u_N(t)$. Since $U_N(f)$ is a narrow bandpass filter centered near $f = \pm 1/T$, only the $l = \pm 1$ terms are non-zero; therefore (2.44) becomes

$$H_N^m(f) = \frac{1}{4|A_N|} \left[U_N\left(f + \frac{1}{T}\right) \int_{-\infty}^{\infty} q_N'(t)q_N(t - mT)e^{j2\pi(f+1/T)t}dt \right.$$

$$\left. + U_N\left(f - \frac{1}{T}\right) \int_{-\infty}^{\infty} q_N'(t)q_N(t - mT)e^{j2\pi(f-1/T)t}dt \right] \quad (2.45)$$

Also, since $U_N(f)$ is designed to be narrow band, and $q_N(t)$ is a pulse which has a duration that is short compared with the jitter, the integral in (2.45) can be considered to be a constant over the band of $U_N(f)$. Therefore the dependence of $H_N(f)$ on m is removed, and (2.45) simplifies to

$$H_N(f) = \frac{U_N\left(f + \frac{1}{T}\right) + U_N\left(f - \frac{1}{T}\right)}{2\left|U_N\left(\frac{1}{T}\right)\right|} \quad \text{for } |f| < \frac{1}{T} \quad (2.46)$$

The jitter transfer function $H_N(f)$ given by (2.46) is the normalized low-pass equivalent of the bandpass filter $U_N(f)$ at frequency $1/T$ [2.21]. Using the same arguments, all second terms of the convolutions in (2.39) can be described by $h_N[kT]$ with transform given by (2.46).

Thus we have derived a jitter transfer function that is dependent only on the characteristic of the timing extraction circuit's narrow bandpass filter. Next, we will show the validity of (2.46) experimentally for a *surface acoustic wave* (SAW) bandpass filter timing extraction circuit [2.22]. We will also show experimentally that the jitter transfer function can be linear over the range of input jitter that we would expect in a well-designed transmission system.

Figure 2.8 shows the measured transfer function $U_N(f)$ of a resonator-type SAW filter timing extraction circuit inside a fiber optic regenerator [2.22]. This bandpass filter has center frequency 274.315 MHz and bandwidth 80 kHz. The filter's passband is smooth and has a linear argument of slope 2.4°/kHz. Using (2.46), we calculated the jitter transfer function for $1/T = 274.315$ Mb/s (i.e., the center frequency of the SAW filter). The calculated jitter transfer function is shown at the top of Figure 2.9a. Note that $|H_N(f)| \leq 1$ (0 dB) for all f. Should the $|H_N(f)| > 1$ for any f, we say that the jitter transfer function exhibits jitter peaking. In Chapter 3, we will see that jitter peaking must be avoided to control jitter accumulation. Also using (2.46), we calculated the jitter transfer function for $1/T$, 20 kb/s from the center frequency of the SAW filter (i.e., 274.335 Mb/s). This calculated jitter transfer function is shown at the top of Figure 2.9b. Note that $H_N(f)$ of the regenerator with a mistuned retiming filter is significantly different from $H_N(f)$ of the properly tuned filter. To verify (2.46), we used the measurement procedure discussed below to measure the jitter transfer function of this regenerator at both bauds of 274.315

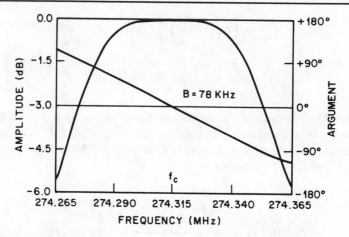

Fig. 2.8 Transfer of a SAW bandpass filter

Mb/s and 274.335 Mb/s. The measured jitter transfer functions are shown at the bottom of Figure 2.9a and 2.9b, respectively. When compared, the calculated and measured results show close agreement. This agreement shows that the jitter transfer function can be dependent only on the characteristic of the bandpass filter.

We measure the jitter transfer function of a regenerator by using a data signal that is phase modulated with a sinusoidal jitter of

$$e_{N-1}(t) = K_i \sin 2\pi f_{pm} t \qquad (2.47)$$

where K_i is the amplitude of the jitter and f_{pm} is the jitter frequency [2.20].

We generate this input data signal by phase modulating a stable reference clock at the baud, with $e_{N-1}(t)$ given by (2.47). (See Figure 2.10.) This clock is then used to clock out a shift register pattern generator whose output bit pattern is sent optically to the regenerator under test. The signal received by the regenerator goes through the timing extraction circuit that outputs

$$s_N(t) = |A_N| \sin \left(\frac{2\pi}{T} t + \frac{2\pi}{T} e_N(t) \right) \qquad (2.48)$$

where $e_N(t) = (e_{iN}(t) + e_{N-1}(t)) * h_N(t)$ from (2.41); $e_N(t)$ is detected by mixing $s_N(t)$ with the reference clock. An important requirement of this measurement procedure is that the reference clock and the phase modulated clock must be frequency locked. This can be done by synthesizing each clock from a single reference source. If $e_{N-1}(t) \gg e_{iN}(t)$, the detected output jitter is

$$e_N(t) = e_{N-1}(t) * h_N(t) \qquad (2.49)$$

where the output jitter is

$$e_N(t) = K_o(f_{pm}) \sin [2\pi f_{pm}t + \theta(f_{pm})] \qquad (2.50)$$

with $K_o(f_{pm})$ being the magnitude of the output jitter, and $\theta(f_{pm})$ the phase difference between the input and output jitter. Using a spectrum analyzer and a phase detector and sweeping the phase modulation frequency f_{pm}, the magnitude and argument of jitter transfer function is measured, where

$$|H_N(f_{pm})| = \frac{K_o(f_{pm})}{K_i} \qquad (2.51\,a)$$

and

$$\arg H_N(f_{pm}) = \theta(f_{pm}) \qquad (2.51\,b)$$

With this measurement technique we should be able to resolve $H_N(f_{pm})$ to 0.01 dB in magnitude and 0.1 degree in argument over a wide range of f_{pm}.

In Figure 2.11, we show results of jitter transfer function measurements made on a fiber optic regenerator with input sinusoidal jitter ranging from 40° to 200° peak-to-peak. Since the curve for each level of input jitter was identical, the timing extraction circuit operates linearly on the input jitter over a wide range.

Since we have shown that (2.46) is experimentally valid, we will use (2.46) to calculate $H_N(f)$ for various analytical models of timing extraction filters. Using (2.46), we can study the effect filter anomalies such as mistuning, passband ripple, and passband asymmetries have on the jitter transfer function. As our first example, we will consider a first-order bandpass filter. A first order bandpass filter has a transfer function given by

$$U_N(f) = \frac{1}{1 + j\dfrac{(f - f_c)}{B}} \quad \text{for } f > 0$$

$$= \frac{1}{1 + j\dfrac{(f + f_c)}{B}} \quad \text{for } f < 0 \qquad (2.52)$$

where B is the full width, half maximum (FWHM) bandwidth and f_c is the center frequency. The magnitude and argument of $U_N(f)$ is graphed in Figure 2.12. Note that f_c is not necessarily tuned exactly to $1/T$, as manufacturing tolerance and filter aging can result in mistuning. Defining a fractional mistuning factor as

$$\eta = \frac{\dfrac{1}{T} - f_c}{B} \qquad (2.53)$$

Comparison of a Calculated and a Measured
Jitter Transfer Function

Measured

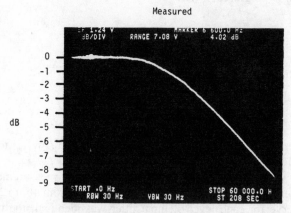

Frequency - (KHz)

Calculated

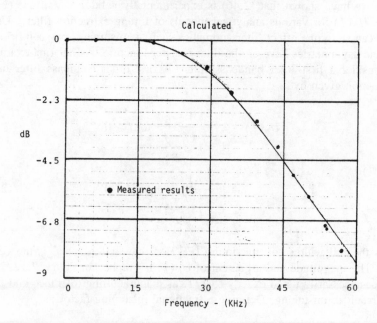

Frequency - (KHz)

Fig. 2.9a Comparison of a calculated and a measured jitter transfer function

Comparison of a Calculated and a Measured
Jitter Transfer Function for a Mistuned SAW Filter

$(\eta = .5)$

Measured

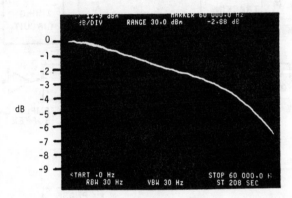

Frequency - (KHz)

Calculated

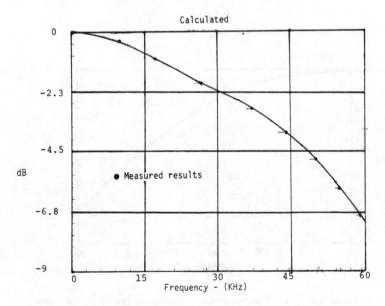

Fig. 2.9b Comparison of a calculated and a measured jitter transfer function for a mistuned SAW filter

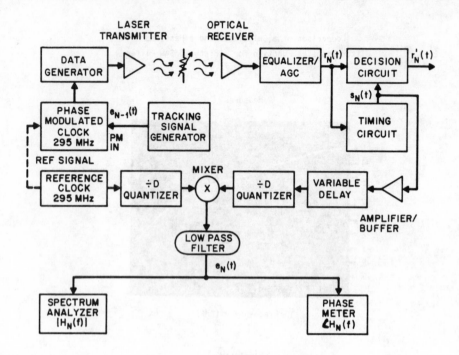

Fig. 2.10 Jitter transfer function measurement configuration

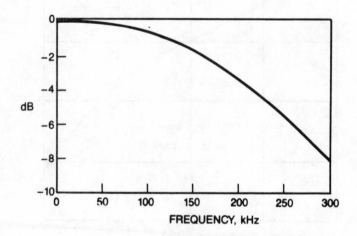

Fig. 2.11 Results of jitter transfer function measurements made on a fiber optic regenerator

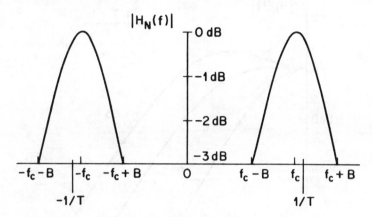

Fig. 2.12 Transfer function of a first-order bandpass filter

and substituting (2.53) and (2.52) into (2.46), we calculate the jitter transfer function for the first-order timing extraction filter as

$$H_N(f) = (1 - \eta^2) \frac{1 + j\dfrac{f}{B}}{\left(1 + j\dfrac{f}{B}\right)^2 - \eta^2} \tag{2.54}$$

In Figure 2.13 we have graphed $|H_N(f)|$ given by (2.54) for various values of η. Figure 2.13, shows that $|H_N(f)|$ does not exhibit jitter peaking unless the filter mistuning is severe ($\eta \geq 0.5$). To manufacture line regenerators that do not exhibit jitter peaking, the bandpass filter must be tuned closely to $1/T$ for the life of the system. Therefore, filter mistuning because of aging must be minimized to avoid a condition where increased jitter accumulation occurs during the life of the system because of an onset of jitter peaking in the regenerators. The requirement of minimized filter mistuning suggests a large filter bandwidth so that jitter peaking does not occur as the filter mistunes because of aging or manufacturing variation. On the other hand, the filter bandwidth should be minimized to filter as much jitter from around the baud component as possible. Therefore, jitter considerations should be used to determine the bandwidth of the timing extraction bandpass filter. Note that with zero mistuning ($f_c = 1/T$, $\eta = 0$)

$$H_N(f) = \frac{1}{1 + j\dfrac{f}{B}} \tag{2.55}$$

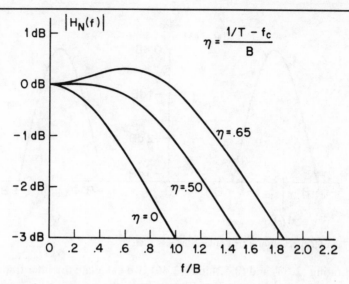

Fig. 2.13 Jitter transfer function of a first-order bandpass filter with various amounts of mistuning

exhibits no jitter peaking. Equation (2.55) is the jitter transfer function used by Bryne in his classic paper on jitter accumulation [2.17]. Because of this, most jitter analyses use the jitter transfer function given by (2.55). Since jitter peaking is not a possibility with the perfectly tuned first-order filter, many jitter analyses fall short of predicting accurate jitter accumulation.

A more realistic bandpass filter for use in analyzing of digital transmission systems is a second-order bandpass filter. For a second-order filter, any passband ripple (in addition to filter mistuning) can result in jitter peaking in the jitter transfer function. A second-order bandpass filter has a transfer function given by

$$U_N(f) = \cfrac{1}{1 + j2\dfrac{\xi}{B}(f - f_c) - \left(\dfrac{f - f_c}{B}\right)^2} \text{ for } f > 0$$

$$= \cfrac{1}{1 + j2\dfrac{\xi}{B}(f + f_c) - \left(\dfrac{f + f_c}{B}\right)^2} \text{ for } f < 0 \qquad (2.56)$$

where ξ is the damping factor. $U_N(f)$ given by (2.56) is graphed in Figure 2.14 for various values of ξ. Note that for values of $|\xi| < 1/\sqrt{2}$ the passband of the filter has ripples (i.e., a non-smooth passband). For the sake of this illustration, we will assume zero mistuning, (i.e., $f_c = 1/T$), and calculate the jitter transfer function using (2.46) as

$$H_N(f) = \frac{1}{\sqrt{\left(1 - \dfrac{f^2}{B^2}\right)^2 - 4\dfrac{\xi^2}{B^2}f^2}} \; \frac{1}{1 + j2\dfrac{\xi}{B}f - \left(\dfrac{f}{B}\right)^2} \tag{2.57}$$

In Figure 2.15 we graph $H_N(f)$ given by (2.57) for various values of ξ. The results show that if there is ripple in the passband of the bandpass filter, then there will be jitter peaking. Therefore, bandpass filters for timing extraction must be designed with minimal passband ripple [2.23].

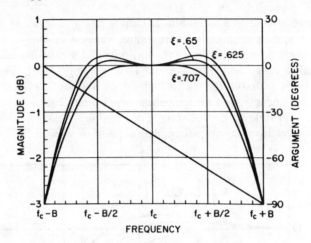

Fig. 2.14 Transfer function of second-order bandpass filter

In the previous two examples, the bandpass filters were assumed to be symmetric around their center frequency. This rarely is the case in practice. Filter asymmetrics can be analyzed using (2.46); however, a graphical construction of $|H_N(f)|$ from the bandpass filter transfer function can also be used. In Figure 2.16, Chamzas [2.21] illustrates the effect filter asymmetry has on $H_N(f)$ for several commonly manufactured bandpass filter shapes. Figure 2.16 shows that filter asymmetries can lead to significant jitter peaking.

In the next chapter, we will show that to control jitter accumulation, jitter peaking must be minimized. To minimize jitter peaking, the bandpass filter of the timing extraction circuit must be designed in such a way that it is closely tuned to the baud frequency for the life of the system, has a ripple free passband, and is symmetric about its center frequency. For high-speed fiber optic systems, there are three alternatives for timing extraction bandpass filters: passive lumped element filters, *phase locked loops* (PLL) [2.18, 2.23] and surface acoustic wave (SAW) filters [2.22].

Both passive lumped element filters and PLLs have serious problems when used in high-speed fiber optic systems. At high frequencies, parasitic capacitance and inductance seriously affects the designed shape of passive lumped element band-pass filters. A circuit designed to be a second-order *resistor, conductor, capacitor* (RLC) circuit at high frequencies can, with stray capacitance and inductance, turn out to be a third-, fourth-, or even a fifth-order filter. This can result in mistuning, passband ripple, and passband asymmetries in the filter shape that would lead to jitter peaking.

PLLs are also difficult to manufacture for use in long-haul, high-speed fiber optic systems. At high frequencies (greater than 200 MHz), the loop's behavior is somewhat unpredictable as a second-order loop is rarely obtained in practice [2.24]. Higher order loops are then difficult to control because of temperature and aging. Also high frequency, higher order (3rd, 4th, 5th) PLLs have the added difficulties of limited pull-in range and injection and false locking [2.25]. These difficulties make SAW filters desirable for use in high-speed fiber optic systems. SAW filters change in a predictable way with temperature and time. They can be designed and manufactured with little or no measured jitter peaking, making them the most desirable timing filter from a jitter accumulation point of view.

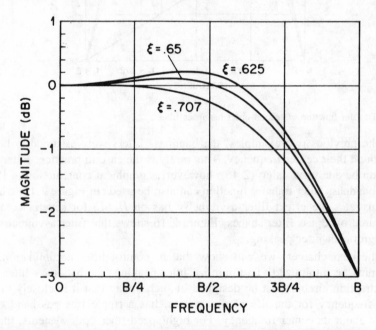

Fig. 2.15 Jitter transfer function of second-order bandpass filter

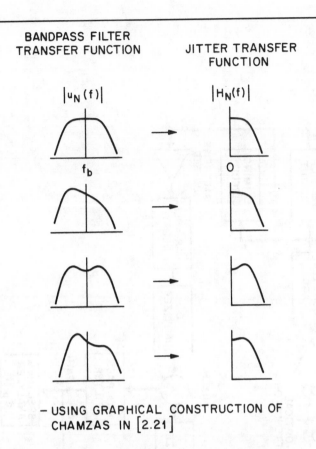

BANDPASS FILTER
TRANSFER FUNCTION

JITTER TRANSFER
FUNCTION

$|u_N(f)|$

$|H_N(f)|$

f_b

0

− USING GRAPHICAL CONSTRUCTION OF
CHAMZAS IN [2.21]

Fig. 2.16 Jitter transfer functions of bandpass filters having passband assymetry

2.6 THE LINEAR SHIFT-INVARIANT JITTER MODEL

Thus far in this chapter, we have shown that a timing signal can be extracted from a cyclostationary received optical signal. We derived a linear, shift-invariant jitter model of the regenerator and discussed the impact on timing circuit design. In this section we will show that the previously derived linear jitter model is experimentally valid for fiber optic regenerators [2.25]. We will do this by measuring the jitter on the timing signal. We will use the configuration shown in Figure 2.17 to measure the jitter on $s_N(t)$; $s_N(t)$ is passed through a quantizer-divider circuit that removes only amplitude variations, and divides the frequency of $s_N(t)$ by a constant D to increase the range of peak jitter that can be measured.

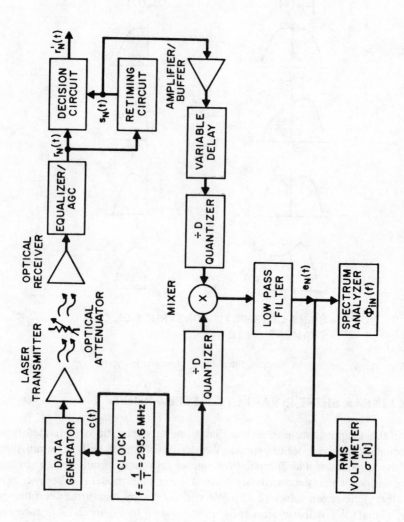

Fig. 2.17 Jitter measurement configuration

We also pass the reference clock signal that clocked out the unjittered data signal through another quantizer-divider circuit. The phase of reference clock relative to $s_N(t)$ is adjusted using an adjustable delay line so that they are in quadrature. The two clocks are then multiplied using a mixer. The mixer output is

$$
\begin{aligned}
m(t) &= K_p \cos \frac{2\pi t}{TD} \sin \left[\frac{2\pi t}{TD} + \frac{2\pi e_N(t)}{TD} \right] \\
&= \frac{K_p}{2} \sin \frac{2\pi e_N(t)}{TD} + \frac{K_p}{2} \sin \left[\frac{4\pi}{TD} + \frac{4\pi e_N(t)}{TD} \right]
\end{aligned}
\tag{2.58}
$$

where K_p is the mixer's phase detection constant in millivolts/degree.

The first term of (2.58) has only low frequency components whereas the second term is centered around $f = 2/DT$. By passing $m(t)$ through a low-pass filter, the output signal of the measurement is

$$
m_f(t) = \frac{K_p}{2} \sin \frac{2\pi e_N(t)}{TD} \approx \frac{K_p \pi}{TD} e_N(t)
\tag{2.59}
$$

where $\sin x \approx x$ approximation was made given that $(2\pi/TD)e_N(t)$ is small. To accommodate large amounts of jitter, D is chosen so that the approximation remains accurate.

Two measurements are made that quantify $e_N(t)$:

1. The rms value, given by $\sigma[N]$, is measured using an rms voltmeter,
2. the power spectrum, given by $\Phi_N(f)$, is measured using a spectrum analyzer.

Using this measurement technique, we will measure each component of $e_N(t)$ derived in the model in Section 2.4.

The pattern-dependent, noise-independent jitter $e_{iN}^P[kT]$ given by (2.40 a) can be shown to be zero mean, with a power spectral density of Φ_{iN}^P, under the assumption of an independent data signal. Using the same assumption and that of a zero mean, white receiver noise process, $e_{iN}^{\eta P}[kT]$ and $e_{iN}^{\eta}[kT]$, can be shown to be zero mean with power spectral densities given by $\Phi_{iN}^{\eta P}$ and Φ_{iN}^{η}, respectively. With these assumptions, the spectrum of the jitter at the output of Nth regenerator when measured with a jitterless incoming signal is

$$
\Phi_N(f) = [\Phi_{iN}^P + \Phi_{iN}^{\eta P} + \Phi_{iN}^{\eta}]|H_N(f)|^2
\tag{2.60}
$$

Using the measurement procedure discussed above, we show in Figure 2.18 $\Phi_N(f)$ measured for a particular fiber optic regenerator for different periodic message sequences. A message sequence with period ν will produce no pattern-dependent jitter (i.e., $e_{iN}^P[kT] + e_{iN}^P[kT] = 0$) if $1/\nu T$ is much larger than the bandwidth of $H_N(f)$. Therefore, using a jitterless 2^7-1 bit shift register pattern as the input message

42

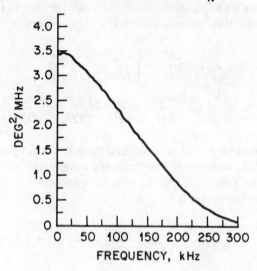

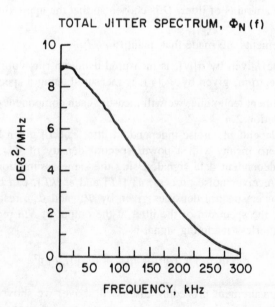

Fig. 2.18 Internally generated jitter spectra for a typical fiber optic regenerator

signal to the regenerator under test, we measure $\Phi_N(f) = \Phi_{iN}^{\eta}|H_N(f)|^2$. At a low frequency ($\approx 10$ kHz) where $|H_N(f)| \approx 1$, we measured $\Phi_{iN}^{\eta} = 3.5$ (degrees)^2/ MHz for a particular regenerator. Recall that Φ_{iN}^{η} is the noise-dependent, pattern-independent jitter generated by a regenerator, and that we used a data pattern with limited pattern variation to measure it. Using a data pattern with maximum pattern variations (e.g., a $2^{23}-1$ bit word), the spectrum measured for an individual regenerator contains both pattern-dependent and noise-dependent jitter (i.e., (2.60)). Again measuring the total jitter power spectral density where $|H(f)| \approx 1$ and subtracting Φ_{iN}^{η} previously measured from the result, we obtain the density of the pattern dependent portion of the jitter ($\Phi_{iN}^{\eta P} + \Phi_{iN}^{P}$) = 5.5 (degrees)^2/MHz.

These results show the difference between pattern-dependent jitter and pattern-independent jitter, and how to measure each. Now we will show experimentally that the pattern-dependent jitter is of two-types: $\Phi_{iN}^{\eta P}$ and Φ_{iN}^{P}. Figure 2.19 shows results of rms jitter measurements made at different received input optical signal powers. Since the purpose of the AGC circuit is to hold the output signal level constant, regardless of the input signal power, we can increase the noise on the AGC output by optically attenuating the input signal to the receiver. An increase in output noise power caused by a lower received signal will produce a higher gain in the AGC, thus resulting in a constant power output signal, but containing more amplifier noise power. By measuring the pattern-dependent jitter as a function of input optical power, we can separate $e_{iN}^{\eta P}$ from e_{iN}^{P}.

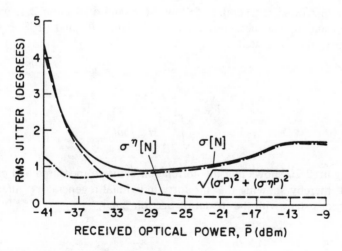

Fig. 2.19 Measured rms jitter vs. received optical power

As expected, $\sigma^\eta[N]$ decreased significantly as the optical signal power increased. This experimentally confirms the derivation in Section 2.4, showing that noise input to the timing extractor produces a noise-dependent jitter on $s_N(t)$. The pattern-dependent jitter also decreased as the optical power increased. This shows the existence of a pattern- and noise-dependent jitter, $e^{\eta P}[kT]$. As the optical power increased, $e^{\eta P}[kT] \to 0$ and the remaining jitter measured was $\sigma^P[N]$, the rms of $\Phi_{iN}^P[kT]$; $\sigma^P[N]$ shown in Figure 2.17 did not significantly change over the entire dynamic range of the receiver. However, it did increase at high input powers, where severe pulse distortion of the received signal caused receiver overload.

Thus we have shown experimentally the existence of the three types of jitter derived in the model of Section 2.3 and showed a procedure to measure each type. In the next chapter, we will cascade forty regenerators and study how each type of jitter accumulates. To set the stage for the next chapter, we will show experimentally that the forty regenerators to be used for the accumulation measurements behave according to the model presented in Section 2.3.

We measured the pattern-dependent and pattern-independent jitter generated by each of the forty regenerators. The input optical power to each regenerator was -30.0 dBm; thus, the pattern-dependent jitter measured was composed of both noise-dependent and noise-independent components. We measured $\sigma^\eta[N]$ and Φ_{iN}^η as well as $\sqrt{(\sigma^{\eta P}[N])^2 + (\sigma^P[N])^2}$ and $(\Phi_{iN}^{\eta P} + \Phi_{iN}^P)$ for forty individual regenerators. We will also measure $H_N(f)$ of each regenerator.

To show that these regenerators follow the model of Section 2.4, we plot in Figures 2.20 a and b, the measured vs. the calculated Φ_{iN}^η and $(\Phi_{iN}^{\eta P} + \Phi_{iN}^P)$, respectively, where the calculated

$$\Phi_{iN} = \frac{\sigma^2[N]}{\displaystyle\int_{-1/T}^{1/T} H_N(f)df} \tag{2.61}$$

Examining Figure 2.20 shows that the correlation between measured and calculated Φ_{iN} was good, thereby showing that the forty individual regenerators' jitter characteristics were consistent with the model of Section 2.4.

2.7 REMARKS

In this chapter, we explored the jitter introduced by line regenerators. We showed that a timing signal can be extracted from a cyclostationary received optical signal. However, the timing extraction process was shown to generate jitter. We then showed experimentally that fiber optic regenerators with SAW filter timing circuits could be modeled using the linear shift-invariant jitter model. The regenerators were shown

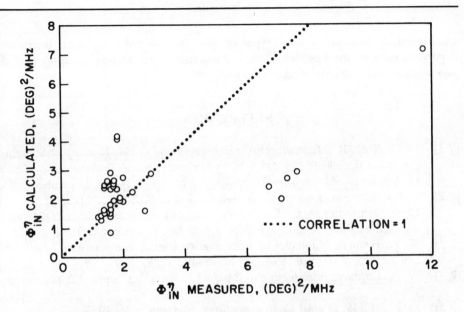

Fig. 2.20a Correlation between measured and calculated noise-dependent jitter spectral density

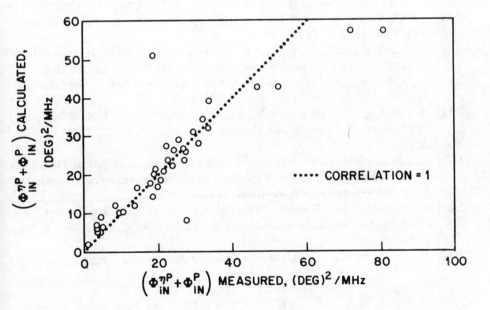

Fig. 2.20b Correlation between measured and calculated pattern-dependent jitter spectral density

experimentally to generate three types of jitter, all filtered by a linear jitter filter dependent only on the bandpass filter characteristic. We are now ready to cascade regenerators and study jitter accumulation.

REFERENCES

[2.1] D. G. Ross, R. M. Paski, D. G. Ehrenberg and G. M. Homsey, "A Highly Integrated Regenerator of 295.6 Mb/s Undersea Optical Transmission," *IEEE/ OSA Journal of Lightwave Technology*, Vol. LT-2, No. 6, December 1984.

[2.2] F. Bosch, G. M. Palmer, C. O. Sallada and C. B. Swan, "Compact 1.3 μm Laser Transmitter for the SL Undersea Lightwave System," *IEEE/OSA Journal of Lightwave Technology*, Vol. LT-2, No. 6, December 1984.

[2.3] W. R. Bennett, "Statistics of Regenerative Digital Transmission," *Bell System Technical Journal*, Vol. 37, November 1958.

[2.4] W. A. Gardner, *Introduction to Random Processes*, Chapter 12, Macmillan, 1986.

[2.5] J. G. Proakis, *Digital Communications*, McGraw-Hill, 1983.

[2.6] S. W. Golomb, *Shift Register Sequences*, Aegean Park Press, 1982.

[2.7] H. Nyquist, "Certain Factors Affecting Telegraph Speed," *Bell System Technical Journal*, Vol. 3, April 1924.

[2.8] L. E. Franks and J. P. Bubrouski, "Statistical Properties of Timing Jitter in a PAM Timing Recovery Scheme," *IEEE Transactions on Communications*, Vol. Com-22, No. 7, July 1974.

[2.9] U. Mengali and G. Perani, "Jitter Accumulation in PAM Systems," *IEEE Transactions on Communications*, Vol. Com-28, No. 8, August 1980.

[2.10] E. O. Sunde, "Self-timing Regenerative Repeaters," *Bell System Technical Journal*, Vol. 36, July 1951, pp. 891–938.

[2.11] A. Luvison, G. Pirani and U. Mengali, "A Simple Timing Circuit for Optical PCM Repeaters," *Optical and Quantum Electronics*, Vol. 13, 1981, pp. 309–322.

[2.12] G. L. Cariolaro and F. Todero, "A General Spectral Analysis of Time Jitter Produced in a Regenerative Repeater," *IEEE Transactions on Communications*, Vol. Com-25, No. 4, April 1977.

[2.13] S. Pupolin and C. Tomasi, "Spectral Analysis of Line Regenerator Time Jitter," *IEEE Transactions on Communications*, Vol. Com-32, No. 5, May 1984.

[2.14] M. D. Greenberg, *Foundations of Applied Mathematics*, Prentice-Hall, NJ 1978.

[2.15] A. Papoulis, *Signal Analysis*, McGraw-Hill, 1977, p. 24.

[2.16] O. Munoz-Rodriguez and K. W. Cattermole, "Time Jitter in Self-Timed Regenerative Repeaters with Correlated Transmitted Symbols," *Electronic Circuits and Systems*, Vol. 3, No. 3, May 1979.

[2.17] C. J. Byrne, B. J. Karafin, and D. B. Robinson. "Systematic Jitter in a Chain of Digital Regenerators," *Bell System Technical Journal*, November 1963, pp. 2679–2714.

[2.18] E. Roza, "Analysis of Phase-Locked Timing Extraction Circuits for Pulse Code Modulation," *IEEE Transactions on Communications*, Vol. Com-22, September 1974.

[2.19] Y. Takasaki, "Timing Extraction in Baseband Pulse Transmission," *IEEE Transactions on Communications*, Vol. Com-20, No. 5, October 1972.

[2.20] P. R. Trischitta, P. Sannuti, "The Validity of the Linear, Shift-Invariant Model of Jitter for a Fiber Optic Regenerator," *IEEE Transactions on Communication*, 1988.

[2.21] R. L. Rosenberg, C. Chamzas and D. A. Fishman, "Timing Recovery with SAW Transversal Filters in the Regenerators of Undersea Long Haul Fiber Transmission Systems," *IEEE/OSA Journal of Lightwave Technology*, Vol. LF2, No. 6, December 1984.

[2.22] R. L. Rosenberg, D. G. Ross, P. R. Trischitta, C. B. Armitage and D. A. Fishman, "Optical Fiber Repeated Transmission Systems Utilizing SAW Filters," *IEEE Transactions on Sonics and Ultrasonics*, Vol. 30, No. 3, May 1983.

[2.23] D. L. Duttweiler, "The Jitter Performer of Phased-Locked Loop Extracting Timing from Baseband Data Waveforms," *Bell System Technical Journal*, Vol. 55, No. 1, January 1976.

[2.24] M. W. Hall, "Unpublished Work on High Frequency Phase Locked Loops for Timing Extraction," AT&T Bell Laboratories, 1983.

[2.25] P. K. Runge, "Phase Locked Loops with Signal Injection for Increased Pull-In Range and Reduced Output Phase Jitter," *IEEE Transactions on Communications*, Vol. Com-24, 1976.

Chapter 3
Jitter Accumulation in Cascaded Regenerators

In this chapter we will study how jitter accumulates when line regenerators are cascaded. Starting with the jitter model of an individual regenerator from Chapter 2, we will derive a jitter accumulation model for a chain of nonidentical regenerators. To test this model, we will cascade 40 fiber optic regenerators, measure the jitter accumulation, and compare measured jitter accumulation with model calculations [3.1]. After showing that the jitter accumulation model is experimentally valid, we will turn our attention to experimental and analytical techniques that can be used to assure proper design and manufacture of regenerators from a jitter accumulation standpoint. We will discuss the circulating loop jitter accumulation simulation technique and analytical jitter accumulation models, assuming identical and statistically distributed cascaded regenerators.

3.1 JITTER ACCUMULATION MODELS AND MEASUREMENTS

In Figure 3.1, we show the jitter model derived in Chapter 2 for the *Nth* regenerator in a chain of cascaded regenerators. Accumulated jitter from the previous $N - 1$ regenerators is passed through a linear jitter filter, along with the jitter internally generated by the regenerator. The output jitter $e_N[nT]$ is the filtered sum of the input and internally generated jitter where the following linear, shift-invariant expression holds

$$e_N[nT] = (e_{N-1}[nT] + e_{iN}[nT]) * h_N[nT] \qquad (3.1)$$

To study jitter accumulation we must separate $e_{iN}[nT]$ into two types: random and systematic. Random jitter is defined as the jitter that is uncorrelated with the jitter generated by the other regenerators in the chain. Systematic jitter, on the other hand, is defined as the jitter that is completely correlated with the jitter generated by the other regenerators in the chain. Pattern-dependent jitter is a source of sys-

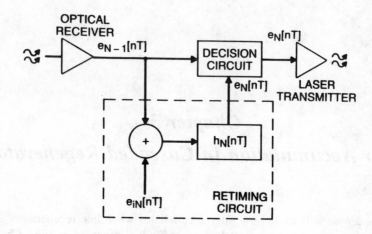

Fig. 3.1 Jitter model of a fiber optic regenerator

tematic jitter because the same bit pattern passes through each regenerator in the chain. However, not all pattern-dependent jitter generated by a regenerator accumulates systematically. In Chapter 2, we showed that a portion of the pattern-dependent jitter may also be noise-dependent, thereby reducing its correlation with pattern-dependent jitter from other regenerators. Therefore, only a portion of the pattern-dependent jitter generated by each regenerator accumulates systematically, and the remaining portion accumulates randomly. We will determine empirically the portion of pattern-dependent jitter generated by each regenerator that accumulates as systematic jitter, and the portion that accumulates as random jitter for a particular set of fiber optic regenerators.

We model separately the accumulation of random and systematic jitter. For a chain of N cascaded nonidentical regenerators (Figure 3.2) the random jitter at the end of the chain using (3.1) is

$$e_N^R[nT] = e_{i1}^R[nT] * h_1[nT] * h_2[nT] * \ldots * h_N[nT]$$
$$+ e_{i2}^R[nT] * h_2[nT] * \ldots * h_N[nT]$$
$$\vdots$$
$$+ e_{iN}^R[nT] * h_N[nT] \tag{3.2}$$

The random jitter generated internally by a particular regenerator is by definition uncorrelated with the jitter generated by all other regenerators in the chain. Assuming each $e_{il}^R[nT]$ is a zero-mean, white stochastic process, the power spectrum of the accumulated random jitter is

$$\Phi_N^R(f) = \Phi_{i1}^R \prod_{l=1}^{N} |H_l(f)|^2 + \Phi_{i2}^R \prod_{l=2}^{N} |H_l(f)|^2$$

$$+ \ldots + \Phi_{iN}^R |H_N(f)|^2 \quad \text{for} \quad |f| < \frac{1}{T} \tag{3.3}$$

where Φ_{il}^R is the constant random jitter power spectral density of $e_{il}^R[nT]$. The rms of the accumulated random jitter is

$$\sigma_R[N] = \sqrt{\int_{-1/T}^{1/T} \Phi_N^R(f) df} \tag{3.4}$$

The systematic jitter at the end of a chain of N cascaded nonidentical regenerators (Figure 3.2) using (3.1) is

$$e_N^S[nT] = e_{i1}^S[nT] * h_1[nT] * h_2[nT] * \ldots * h_N[nT]$$

$$+ e_{i2}^S[nT] * h_2[nT] * \ldots * h_N[nT]$$

$$\vdots$$

$$+ e_{iN}^S[nT] * h_N[nT] \tag{3.5}$$

The systematic jitter generated internally by a particular regenerator is, by definition, completely correlated with the systematic jitter generated by the other regenerators in the chain. Assuming each $e_{il}^S[nT]$ is a zero-mean, white stochastic process, the power spectrum of the accumulated systematic jitter is

$$\Phi_N^S(f) = \left| \sqrt{\Phi_{i1}^S} \prod_{l=1}^{N} H_l(f) + \sqrt{\Phi_{i2}^S} \prod_{l=2}^{N} H_l(f) + \ldots \right.$$

$$\left. + \sqrt{\Phi_{iN}^S} H_N(f) \right|^2 \quad \text{for} \quad |f| < \frac{1}{T} \tag{3.6}$$

where Φ_{il}^S is the constant systematic jitter power spectral density of $e_{il}^S[nT]$. The rms of the accumulated systematic jitter is

$$\sigma_S[N] = \sqrt{\int_{-1/T}^{1/T} \Phi_N^S(f) df} \tag{3.7}$$

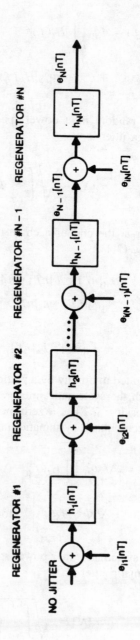

Fig. 3.2 Jitter accumulation model for a chain of regenerators

Since random and systematic jitter are uncorrelated with each other, the total jitter power spectrum $\Phi_N(f)$, and the total rms jitter $\sigma[N]$, at the output of the Nth cascaded regenerator, are

$$\Phi_N(f) = \Phi_N^R(f) + \Phi_N^S(f) \tag{3.8}$$

and

$$\sigma[N] = \sqrt{\sigma_R^2[N] + \sigma_S^2[N]} \tag{3.9}$$

Now that we have derived an expression for the power spectrum and the rms value of the accumulated jitter at the output of the Nth cascaded nonidentical regenerator, we shall exercise this model with an example using fiber optic regenerators.

We have measured the jitter characteristics of forty individual fiber optic regenerators and used their measured values as the parameters in the jitter accumulation model just derived [3.1].

To characterize the jitter contribution of a particular regenerator l, three parameters must be measured: the internally-generated, pattern-dependent, jitter power spectral density ($\Phi_{il}^{PD} = \Phi_{il}^{\eta P} + \Phi_{il}^{P}$); the internally generated, noise-dependent, jitter power spectral density Φ_{il}^{η}; and the jitter transfer function $H_l(f)$. These three parameters were measured on forty individual fiber optic regenerators [3.2] using the measurement techniques discussed in Chapter 2. In Table 3.1, we show the measured Φ_{il}^{PD} and Φ_{il}^{η} of these forty regenerators. Note that for most regenerators, Φ_{il}^{PD} was significantly larger than Φ_{il}^{η}.

To apply these measured parameters to the model discussed previously, we make the following assumptions. Since noise-dependent jitter is uncorrelated with that generated by the other regenerators in the chain, this jitter will accumulate randomly. Pattern-dependent jitter, on the other hand, is composed of two types, and is not necessarily completely correlated with the pattern-dependent jitter generated by the other regenerators in the chain. Only the correlated portion of the pattern-dependent jitter will accumulate as systematic jitter with the uncorrelated portion accumulating as random jitter. We assign the portion of pattern-dependent jitter generated by the lth regenerator that accumulates systematically the constant α_l. Figure 3.3 shows this assignment graphically. However, α_l is unknown when characterizing an individual regenerator; therefore we will assume, for the time being, that the measured pattern-dependent jitter is systematic jitter, (i.e., ($\alpha_l = 1$)).

The magnitude and the argument of $H_l(f)$ were measured on these forty regenerators using the measurement technique discussed in Chapter 2. Figure 3.4 shows the mean and the mean plus and minus one standard deviation of the forty individually measured jitter transfer function magnitudes. No measurable jitter peaking was observed on any regenerator. These forty jitter transfer functions had -3 dB bandwidth ranging from 160 kHz to 220 kHz, and linear argument of slope $0.612°/$kHz.

Table 3.1 Measured Jitter Parameters of 40 Fiber Optic Regenerators

Regenerator, l	Φ_{il}^{η} $(degrees)^2/MHz$	Φ_{il}^{PD} $(degrees)^2/MHz$
1	1.3	52.2
2	6.7	22.1
3	2.3	5.0
4	1.3	24.1
5	1.9	33.3
6	1.9	22.6
7	2.0	18.5
8	1.5	25.5
9	1.6	20.6
10	1.5	17.9
11	1.2	8.5
12	1.6	14.3
13	1.3	26.7
14	1.8	33.6
15	1.7	1.2
16	2.7	30.8
17	1.4	18.9
18	1.6	46.6
19	1.8	31.9
20	1.5	23.9
21	1.4	19.7
22	1.6	24.0
23	2.9	27.0
24	1.6	71.5
25	1.4	21.5
26	1.4	80.5
27	1.6	29.3
28	1.7	33.3
29	1.6	10.6
30	1.7	13.7
31	1.7	27.6
32	1.4	3.7
33	1.7	27.3
34	1.8	18.5
35	7.3	3.8
36	7.6	4.5
37	2.0	5.0
38	11.7	4.6
39	7.1	9.7
40 (always last)	1.4	19.2
Mean	2.5	23.3
Standard Deviation	2.2	16.6

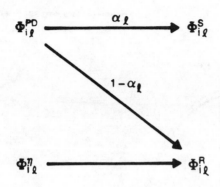

Fig. 3.3 Portion α_l of pattern-dependent jitter that will accumulate as systematic jitter

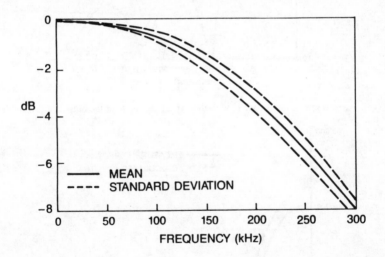

Fig. 3.4 Mean jitter transfer function of forty regenerators along with mean I standard deviation

Inserting these measured individual jitter parameters in the jitter accumulation model, we calculated the jitter accumulation for a chain of such regenerators. Specifically, Φ_{il}^R (Φ_{il}^η) and Φ_{il}^S (Φ_{il}^{PD}) and the corresponding $H_l(f)$ were inserted, in the order shown in Table 3.1, into (3.3) and (3.6) respectively. The total jitter spectrum $\Phi_N(f)$ given by (3.8) was calculated at the output of the sixteenth, twenty-eighth, and fortieth cascaded regenerators. These forty regenerators were then cascaded in the same order, and $\Phi_N(f)$ was measured at the same points. Figures 3.5, 3.6 and 3.7 show the calculated and the measured $\Phi_N(f)$. Examining the calculated $\Phi_N(f)$,

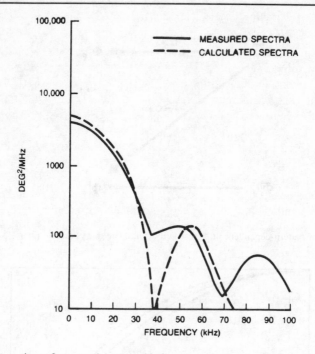

Fig. 3.5 Comparison of measured spectra with the expected spectra at the output of the 16th regenerator

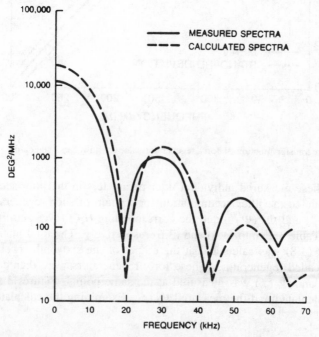

Fig. 3.6 Comparison of measured spectra with the expected spectra at the output of the 28th regenerator

we see that most of the jitter power is near DC with a power density that is increasing roughly as N^2. The spectrum also contains nulls caused by phase cancellations in the vector addition of jitter transfer functions. Both of these spectral characteristics are the result of the assumed systematic accumulation of the pattern-dependent jitter. Random jitter accumulation does not produce nulls in its spectrum, and will have a power density near DC that increases as N.

Comparing the measured and calculated jitter spectra shows agreement in the location of the null frequencies, but the magnitude of the measured spectra and the depth of the nulls were less than what was calculated. Also, as the number of cascaded regenerators increase, the disagreement in magnitude and null depth increases. Since a characteristic of systematic jitter accumulation is a power spectrum with deep nulls and magnitude near DC proportional to N^2, the disagreement can be resolved by assuming more random and less systematic jitter accumulation (e.g., $\alpha_l \neq 1$). With $\alpha_l \neq 1$, the nulls in the calculated spectra would be filled with random jitter and the magnitude of the calculated spectra near DC would be less.

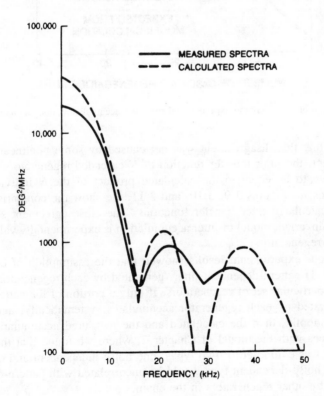

Fig. 3.7 Comparison of measured spectra with the expected spectra at the output of the 40th regenerator

A similar disagreement between calculation and measurement is observed in Figure 3.8, where $\sigma[N]$ given by (3.9) is compated with $\sigma[N]$ measured along the chain of these forty cascaded regenerators. Although near agreement is observed up to ten cascaded regenerators, the measured rms jitter accumulation is considerably less than the calculation afterwards. This disagreement can again be resolved by assuming more random and less systematic jitter accumulation.

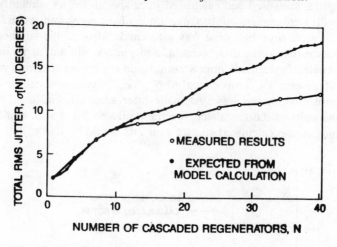

Fig. 3.8 Comparison of measured and expected total rms jitter accumulation along a chain of regenerators

To show that this disagreement was not caused by some nonlinear jitter accumulation effect, the jitter transfer function of N cascaded regenerators was measured and shown to be equal to the calculated product of the N individual jitter transfer functions. In Figures 3.9, 3.10, and 3.11, we show the comparison of the measured and calculated jitter transfer functions. The close agreement shows that the linear, shift-invariant model of jitter accumulation is experimentally valid for this set of cascaded regenerators.

The previous experimental results showed that the assumption of completely correlated ($\alpha_l = 1$) pattern-dependent jitter, generated by each regenerator, was not correct for this particular set of regenerators. If only a portion of the pattern-dependent jitter generated by each regenerator accumulates systematically, and the rest accumulates randomly, then the calculated and the measured accumulation would agree. This agrees with the model of Chapter 2, where we found that the pattern-dependent jitter was composed of two types, one noise-independent and one noise-dependent. The noise-dependent type would be uncorrelated with pattern-dependent jitter generated by other regenerators in the chain.

Using the results of the previous section, we calculated an estimate of the portion of pattern-dependent jitter that accumulated systematically for this set of forty regenerators. Figure 3.12 shows the mean of the α_l's for $l = 1$ to N that makes the calculated jitter accumulation agree with measured accumulation. We see that, as

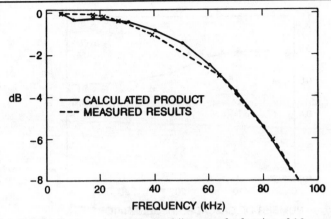

Fig. 3.9 Comparison of measured and calculated jitter transfer function of 16 cascaded regenerators

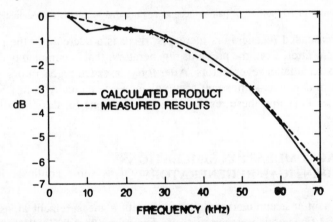

Fig. 3.10 Comparison of measured and calculated jitter transfer function of 28 cascaded regenerators

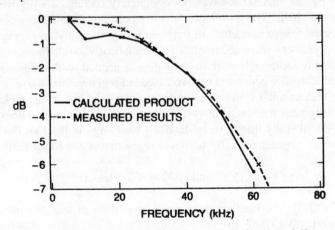

Fig. 3.11 Comparison of measured and calculated jitter transfer function of 40 cascaded regenerators

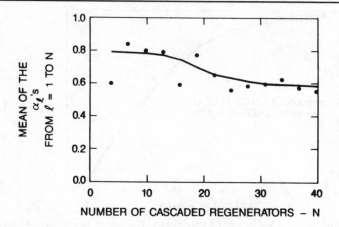

Fig. 3.12 An estimate of the portion of pattern-dependent jitter that accumulates systematically

the number of cascaded regenerators increases, there is a decrease in the portion of pattern-dependent jitter, from the previous regenerators, that is completely correlated with that generated at latter regenerators. After forty cascaded regenerators, the mean portion of pattern-dependent jitter that accumulates systematically is just 56%.

We will see later that these results have ramifications for designing long-haul transmission systems.

3.2 JITTER ACCUMULATION SIMULATIONS ASSUMING IDENTICAL REGENERATORS

By comparing a jitter accumulation calculation with a measurement along a chain of forty cascaded fiber optic regenerators, we determined that the pattern-dependent jitter generated by one regenerator was not completely correlated with that of other regenerators in the chain. The correlation of jitter between regenerators became less as more regenerators were cascaded. In our measurements, only an average of 56% of the pattern-dependent jitter accumulated systematically, with the remaining accumulated randomly. Although more investigation is needed on the cross-correlation of pattern-dependent jitter generated by two cascaded regenerators, these results may impose tighter restrictions on the magnitude of jitter peaking in manufactured regenerators for long-haul transmission systems to avoid exponential random jitter accumulation. We will study this in more detail by studying the jitter accumulation of a chain of identical regenerators. By identical regenerators we mean that

$$H_1(f) = H_l(f) \quad \text{and} \quad \Phi_{il}^R = \Phi_{i1}^R \quad \text{for all } l \in [1,N] \tag{3.10}$$

Using (3.10) in (3.3), the random jitter power spectrum at the output of the *Nth* identical regenerator is [3.3–3.13]

$$\Phi_N^R(f) = \Phi_{i1}^R \sum_{n=1}^{N} |H_1(f)|^{2n} \quad \text{for} \quad |f| < \frac{1}{T} \tag{3.11}$$

$$= \Phi_{i1}^R |H_1(f)|^2 \frac{1 - |H_1(f)|^{2N}}{1 - |H_1(f)|^2} \tag{3.12}$$

We also specialize the systematic jitter accumulation given by (3.6) to a chain of identical cascaded regenerators, so that

$$H_l(f) = H_1(f) \quad \text{and} \quad \Phi_{il}^S = \Phi_{i1}^S \quad \text{for all } l \in [1,N] \tag{3.13}$$

Using (3.13) in (3.6), the systematic jitter power spectrum at the output of the *Nth* identical regenerators [3.3–3.13] is

$$\Phi_N^S(f) = \Phi_{i1}^S \left| \sum_{n=1}^{N} H_1^n(f) \right|^2 \quad \text{for} \quad |f| < \frac{1}{T} \tag{3.14}$$

$$= \Phi_{i1}^S |H_1(f)|^2 \frac{|1 - H_1^N(f)|^2}{|1 - H_1(f)|^2} \tag{3.15}$$

To illustrate the accumulation of random and systematic rms jitter for a chain of identical regenerators, we will assign that $H_1(f)$ of the identical regenerators to be a second-order low-pass filter, e.g.

$$H_1(f) = \frac{B^2}{B^2 + j2\xi Bf - f^2} \quad \text{for} \quad |f| < \frac{1}{T} \tag{3.16}$$

where B is the -3 dB bandwidth and ξ is the damping factor. We saw in Chapter 2 that this $H_1(f)$ can be realized by a second-order PLL, SAW filter, or a second order RLC retiming circuit. The $|H_1(f)|$ is graphed in Figure 3.13 for various amounts of jitter peaking (i.e., maximum $|H_1(f)|/|H_1(0)|$ over all f). Substituting (3.16) into (3.12) and (3.15), we plot in Figures 3.14 and 3.15, $\Phi_N^R(f)$ and $\Phi_N^S(f)$, respectively, for the case where $|H_1(f)|$ has 0 dB of jitter peaking (curve a, Figure 3.13) (e.g., $|H_1(f)| \leq |H_1(0)|$). Comparing Figures 3.14 and 3.15, we see that $\Phi_N^S(f)$ has power near DC that is increasing as N^2, whereas $\Phi_N^R(f)$ has power near DC that is increasing as N. Also, the systematic spectrum has a narrow bandwidth and nulls caused by phase cancellations in the vector addition of the jitter transfer functions. This will limit the accumulation of rms jitter. However, with no jitter peaking, systematic jitter clearly dominates the total jitter accumulation.

Using (3.15) and (3.12) in Figures 3.16 and 3.17, we also plot the normalized rms random jitter accumulation $\sigma_R[N]/\sigma_R[1]$ and the normalized rms systematic jitter accumulation $\sigma_S[N]/\sigma_S[1]$, respectively, *versus* the number of cascaded identical re-

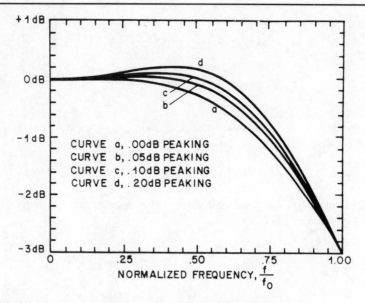

Fig. 3.13 Magnitude of second-order Butterworth jitter transfer function

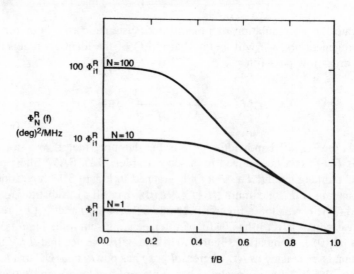

Fig. 3.14 Random jitter spectrum accumulation

generators for an $|H_1(f)|$ with 0 dB to 0.2 dB of jitter peaking (e.g., curves a through d in Figure 3.13).

Comparing Figures 3.16 and 3.17, the systematic rms jitter accumulates roughly as $\sqrt{N}$ and the random rms jitter accumulates roughly as $\sqrt[4]{N}$ when $H_1(f)$ has no jitter peaking. With no jitter peaking, systematic jitter accumulates much more rapidly than random jitter and dominates the total jitter accumulation if $\Phi_{i1}^S \geq \Phi_{i1}^R$.

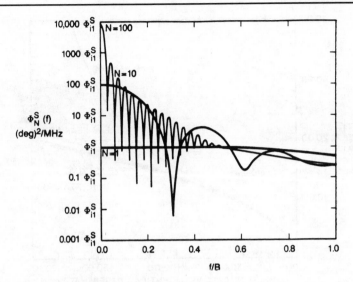

Fig. 3.15 Systematic jitter spectrum accumulation

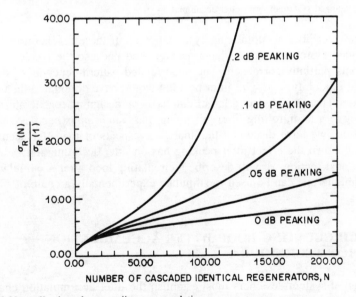

Fig. 3.16 Normalized random rms jitter accumulation

With jitter peaking, the systematic jitter accumulates exponentially, but not as rapidly as the exponential accumulation of random jitter. From this example, clearly with jitter peaking, random jitter should not be ignored since the total jitter accumulation can be dominated by the random jitter accumulation [3.10]. Moreover, in Section 3.1, the regenerators that were cascaded had only an average of 56% of their

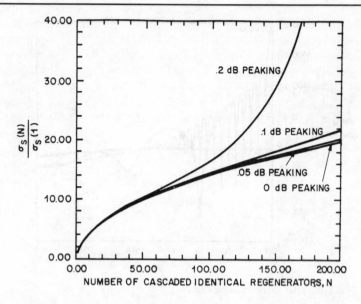

Fig. 3.17 Normalized systematic rms jitter accumulation

pattern-dependent jitter accumulating systematically. If these regenerators had jitter peaking, much more jitter would have accumulated because the pattern-dependent jitter was only partially correlated. The uncorrelated pattern-dependent jitter (44%) accumulated randomly, and with jitter peaking would have dominated the total jitter accumulation. These results put a larger emphasis on designing regenerators without jitter peaking (e.g., controlling filter mistuning, passband ripples, and asymmetries). In this section, we have discussed the analytical aspects of the jitter accumulation models, and shown the effect jitter peaking has on jitter accumulation.

In the next section we will describe a circulating loop jitter accumulation simulation technique that can be used to stimulate experimentally a chain of fiber optic regenerators.

3.3 THE CIRCULATING LOOP JITTER ACCUMULATION SIMULATION TECHNIQUE

Usually, digital system designers must establish the jitter accumulation characteristics of the systems they are developing without the advantage of having many test regenerators. With only a few regenerators and fiber spans, it is difficult to qualify a large, long distance system, since cascading these few provides little experimental information about a long chain of regenerators. If the system designer attempts to qualify system jitter performance based on the analytical jitter accumulation models alone, great care must be taken to assure that the jitter parameters assumed in the analytical jitter accumulation models match those of the manufactured regenerators.

Alternatively, an experimental jitter accumulation simulation technique can be used to simulate the jitter contribution of a particular regenerator (whether in design or in manufacture) including its effect on the jitter accumulation of a long system made up of these regenerators. This would allow for changes to be made to regenerators in manufacture, well before system installation. Various attempts have been made to simulate and study jitter accumulation experimentally. During the development of the Bell System's T1 system, Manley [3.14] characterized the accumulation of jitter in digital regenerators by using an experimental simulation of a long chain of regenerators. One regenerator and a four-track tape recorder were used to simulate a long chain of identical regenerators. While the previous regenerator output was being reproduced from two tracks of the recorder and used as the input to the regenerator, the new output was recorded on the other two tracks. One of each pair of tracks was used for the pulse train and the other for timing information. Manley used this experimental simulation of jitter accumulation along a chain of regenerators to experimentally investigate the effects that filter mistuning, AM to PM conversion, and received pulse shape had on jitter accumulation. These experimental simulations of jitter accumulation were the first such simulations of a chain of regenerators that confirmed the analytical models. Manley's work showed that jitter accumulation could be effectively studied experimentally through simulation of a long chain of regenerators.

However, Manley's experimental technique had severe limitations that would make it impossible to use on today's regenerators, with bit rates in the hundreds of megabits per second. Another simulation technique, also tried during T1 system development, showed more potential for use with fiber optic regenerators than Manley's technique. This experimental technique simulated the jitter accumulation in a chain of identical regenerators by using a single regenerator in a circulating loop with finite delay.

A circulating data loop was first done by Wrathall in 1956 [3.15] when he succeeded in circulating a single digital pulse around a waveguide and regenerated that pulse without error using a regenerator. Carbrey [3.16], together with Byrne, attempted to repeat this circulating loop experiment, using a T1 regenerator and a crystal as a long delay element; but the difficulty in closing the loop without a severe transient clouded the results. Although Kinariwala [3.17] described the potential and the limitations of a circulating loop jitter accumulation experiment, no experimental results were published until [3.18]. We will describe results of circulating loop jitter accumulation simulations using fiber optic regenerators and long single-mode fiber spans and compare these results with jitter accumulation model results.

In the circulating loop jitter accumulation simulation, we are attempting to simulate experimentally, the jitter accumulation along a chain of regenerators. This is done by taking one link of the chain (in particular, and one regenerator and one fiber span), and closing the loop by connecting the output of the regenerator to its input through the fiber span). This technique is illustrated in Figure 3.18. If we record the output jitter $e_1[nT]$ on the extracted timing signal after closing the loop, we can study

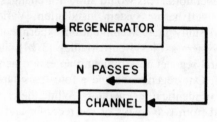

DATA CIRCULATING AROUND CLOSED
LOOP N TIMES WILL SIMULATE DATA
GOING THROUGH N REGENERATORS IN
A CHAIN

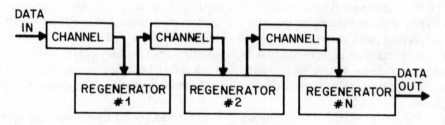

Fig. 3.18 Circulating loop simulation process

the resulting transient waveform in sections. Each section represents the amount of time it takes the data to travel once around the regenerator-fiber loop. The main premise of this experimental technique is that the jitter recorded for each pass of data around the loop simulates the jitter at the output of another regenerator along the chain of identical regenerators.

In Section 3.2, we develop the jitter accumulation model for a chain of identical regenerators. By comparing the loop output with the chain output, an understanding of the analytical basis of the simulation results can be achieved. Applying the linear, shift-invariant jitter model to the loop experimental model, we record the output jitter $e_1[nT]$ for $n = 1$ to K bit periods after the closing of the loop. The output jitter $e_1^{(L)}[nT]$ for the first pass of data around the loop is then

$$e_1^{(L)}[nT] = e_{i1}[nT] * h_1[nT] \quad \text{for} \quad n = 1 \text{ to } K \tag{3.17}$$

where $e_{i1}[nT]$ is the internally generated jitter of the regenerator in the loop, and $h_1[nT]$ is its jitter impulse response. The superscript (L) on $e_1^{(L)}[nT]$ signifies that the output jitter is from the first pass of data around the loop. The integer K is the number of data pulses circulating in the loop. We then calculate the standard deviation for K samples of $e_1^{(L)}[nT]$, i.e.

$$\sigma^L[1] = \sqrt{\frac{1}{K} \sum_{n=1}^{K} (e_1^{(L)}[nT] - \overline{e_1^{(L)}[nT]})^2} \tag{3.18}$$

where $\overline{e_1^{(L)}[nT]}$ is the mean value of $e_1^{(L)}[nT]$ for the $n = 1$ to K interval [3.19]. If (3.18) is equivalent to the rms value $\sigma[1]$ given by (3.9), the output jitter of the first regenerator in a chain of regenerators is simulated by the data going around the loop once. Since this is true for a stationary jitter process in the limit $K \to \infty$, we desire as many bits circulating around the loop as possible, to accurately simulate jitter accumulation.

After the second pass of the data around the loop, the output jitter is

$$e_2^{(L)}[nT] = e_{i1}[nT] * h_1[nT] + e_1^{(L)}[(n - K)T] * h_1[nT]$$

$$\cdot \text{ for } \quad n = K \text{ to } 2K$$

$$= e_{i1}[nT] * h_1[nT] + [e_{i1}[(n - K)T]$$

$$* h_1[(n - K)T]] * h_1[nT] \tag{3.19}$$

with a standard deviation of

$$\sigma^L[2] = \sqrt{\frac{1}{K} \sum_{n=K}^{2K} (e_2^{(L)}[nT] - \overline{e_2^{(L)}[nT]})^2} \tag{3.20}$$

where $\overline{e_2^{(L)}[nT]}$ is the mean value of $e_2^{(L)}[nT]$ for $n = K$ to $2K$. Again, if (3.20) is equivalent to $\sigma[2]$, as given by (3.9), the output jitter of the second regenerator in the chain of identical regenerators is simulated by the now jittered data going around the loop a second time.

Data continues to circulate around the loop and the output jitter is recorded. After N passes of data around the loop, the output jitter is given by

$$e_N^{(L)}[nT] = e_{i1}[nT] * h_1[nT] + e_{N-1}^{(L)}[(n - K)T] * h_1[nT]$$

$$\text{for } \quad n = (N - 1)K \text{ to } NK$$

$$= e_{i1}[nT] * h_1[nT] + e_{i1}[(n - K)T] * h_1[(n - K)T] * h_1[nT]$$

$$+ e_{i1}[(n - 2K)T] * h_1[(n - 2K)T] * h_1[(n - K)T]$$

$$* h_1[nT] + \ldots + e_{i1}[(n - (N - 1)K)T]$$

$$* h_1[(n - (N - 1)K)T] * \ldots * h_1[(n - K)T] * h_1[nT] \tag{3.21}$$

We calculate its standard deviation as

$$\sigma^L[N] = \sqrt{\frac{1}{K} \sum_{n=(N-1)K}^{NK} (e_N^{(L)}[nT] - \overline{e_N^{(L)}[nT]})^2} \tag{3.22}$$

where $\overline{e_N^{(L)}[nT]}$ is the mean value of $e_N^{(L)}[nT]$ for $n = K$ to $(N - 1)K$. If (3.22) is equivalent to $\sigma[N]$, as given by (3.9), the output jitter of the Nth regenerator in a chain of identical regenerators is effectively simulated by the data going around the loop the Nth time.

This equivalence would be true as long as the N convolutions of $h_1[nT]$ in (3.21) does not have a time duration longer than KT. The time duration of N convolutions of $h_1[nT]$ is roughly $N|\phi_p|$ where ϕ_p is the slope of the linear argument of $H_1(f)$ [3.20]. Therefore as soon as

$$N|\phi_P| > KT \tag{3.23}$$

the circulatory loop does not accurately simulate the jitter accumulation along a chain of regenerators and will reach a steady state [3.20–3.24]. However, until this limitation is reached (e.g., $N < KT/|\phi_p|$, the simulation is a valid measurement of jitter accumulation along a chain of N identical regenerators.

Equation (3.21) can be simplified for systematic jitter. With a repetitive pattern circulating around the loop, the systematic jitter generated by the regenerator in the loop is periodic, so that

$$e_{i1}^S[nT] = e_{i1}^S[(n - K)T] \tag{3.24}$$

for all n. This assumes that the repetitive pattern used to start the loop is larger than the pattern captured in the loop. Substituting (3.24) into (3.21) yields the loop model for systematic jitter accumulation as

$$e_N^{S(L)}[nT] = e_{i1}^S[nT] * \left(h_1[nT] + h_1[nT] * h_1[nT] \right.$$

$$\left. + \ldots + h_1[nT]_{N\ times}^{*\cdots*} h_1[nT] \right) \tag{3.25}$$

for $n = (N - 1)K$ to NK.

Equation (3.25) is similar to the systematic jitter accumulation model for a chain of identical regenerators given by (3.5) and is equivalent in the limit as $K \to \infty$. Therefore, as long as K is large such that $h_1[nT]_{N\ times}^{*\cdots*} h_1[nT]$ has time duration less than KT, the circulating loop will simulate cascaded regenerators. We can easily extend this model for more than one regenerator-fiber span in the loop.

Figure 3.19 shows the block diagram of the experimental circulating loop jitter accumulation simulation setup. A stable reference clock at frequency $1/T = 295.6$

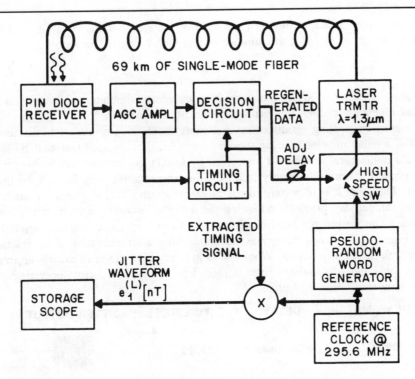

Fig. 3.19 Experimental setup of circulating loop jitter accumulation simulation

MHz is used to clock out a shift register bit with period $2^{23} - 1$ bits. This data signal is passed through a high-speed digital switch that is opened to start the experiment by filling the line with pulses. The output of the digital switch modulates the laser transmitter that is connected to a long length l, here 69 km, of single-mode fiber. The optical pulse pattern is received, a timing signal extracted, and a new signal regenerated by the fiber optic regenerator under test. The regenerated output data signal of the decision circuit is then connected to the second input port of the digital switch. The jitter $e^{(L)}[nT]$ on the extracted timing signal is detected the same way as in the measurement procedures of Chapter 2 (e.g., the timing signal is multiplied in quadrature with the reference clock; $e^{(L)}[nT]$ is then recorded using a transient recorder.

By closing the loop switch rapidly, data is circulated around the regenerator-fiber loop, simulating data going through a chain of identical regenerators spaced l kilometers apart. By starting the transient recorder at the instant the switch is closed, the jitter signal $e^{(L)}[nT]$, is recorded. Since the delay τ_L for one complete pass around the loop is given by

$$\tau_L = \frac{ln_c}{c} \tag{3.26}$$

where n_c is the refractive index of the core of the single-mode fiber ($n_c = 1.453$) and c is the speed of light, the number of bits circulating in the loop is

$$K = \frac{\tau_L}{T} \tag{3.27}$$

Each time duration τ_L of the transient jitter waveform simulates the output jitter of another regenerator along the chain of identical regenerators.

Before presenting results of several simulations, it is necessary to discuss the effect of a delay mismatch at the switch-over point. If τ_L is such that switch closure allows a nonintegral number of bits given by (3.27) to circulate around the loop, a sharp phase discontinuity occurs each time the fractional bit goes through the timing circuit. Therefore it is important to have a stable symbol period T and a τ_L such that K is an integer. By precisely adjusting the delay τ_L around the loop, it is possible to ensure that an integer number of bits, K in number, are circulating around the loop. Figure 3.20 shows these phase discontinuities after each pass of the fractional bit through the timing circuit. When the loop delay is trimmed to exactly an integer of bits, only the accumulated jitter is recorded. This is the correct operating point of the loop simulation technique.

LENGTH MISMATCH	DELAY	τ/T (TIMESLOTS)	JITTER OUTPUT
6 cm SHORT	-300 psec	1/12	
4 cm SHORT	-200 psec	1/24	
2 cm SHORT	-100 psec	1/36	
0 cm	0 psec	0	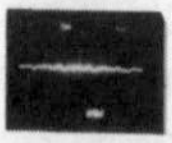
2 cm LONG	+100 psec	1/36	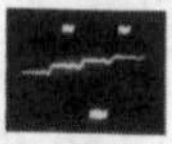
4 cm LONG	+200 psec	1/24	
6 cm LONG	+300 psec	1/12	

PHASE DISCONTINUITIES DUE TO LENGTH MISMATCHES ARE EFFECTIVELY TUNED OUT WHEN LENGTH OF LOOP IS WITHIN ±1 cm (1/72 TIMESLOTS) OF OPTIMUM

Fig. 3.20 Effects of length mismatch on loop output

Figure 3.21 shows a typical loop experiment transient waveform output. Since the delay τ of the loop is known, we calculate the standard deviation of the jitter recorded during the *Nth* pass of data around the loop and graph $\sigma^L[N]$ versus the number of cascaded regenerators simulated.

We simulated the jitter accumulation of a chain of one hundred ninety-one identical regenerators by looping the data signal of one fiber optic regenerator [3.1] through a 69 km length of single-mode fiber [3.25]. With a fiber length $l = 69$ km, the loop delay τ_L was 330 μs. At symbol period $T = 3.3$ ns there were 98,301 bits circulating around this fiber-regenerator loop. The first results were obtained for a regenerator with $|H_1(f)|$ shown in Figure 3.22a, having 0.00 dB of jitter peaking.

Making sure that the loop delay was trimmed to exactly an integer number of bits, the loop was closed and the output jitter transient waveform was recorded. In Figure 3.23a we show an oscilloscope photograph of the output transient waveform. By dividing this transient into 330 μs sections, we calculated $\sigma^L[N]$ for successive passes around the loop. In Figure 3.24, $\sigma^L[N]$ is plotted versus N along with the calculated curve given by (3.9) with identical regenerators. Since no jitter peaking was measured in the regenerator's jitter transfer function, an accumulation of less than the square root of the number of regenerators was expected. Figure 3.24 shows agreement between the measured results and model calculations, as no exponential growth was observed in either the calculations or the simulation.

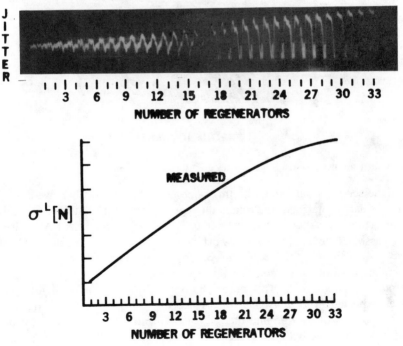

Fig. 3.21 Simulated jitter accumulation of a chain of identical regenerators

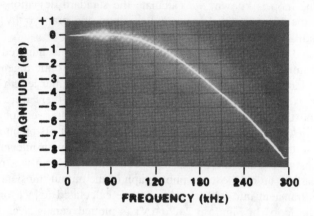

NO MISTUNING — .00 dB JITTER PEAKING

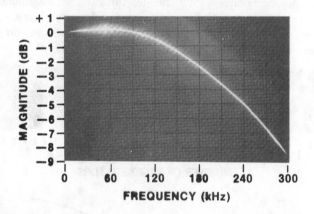

35 kHz MISTUNING — .15 dB JITTER PEAKING

Fig. 3.22 Jitter transfer function of the regenerator used in loop simulation

In the second simulation with this same regenerator and fiber, we changed the baud of the data signal passing through the regenerator, so the data rate was mistuned from the center frequency of the SAW filter. In this simulation the SAW filter (containing passband ripple) [3.10] was mistuned 35 kHz so that a jitter peak of 0.15 dB was present in $|H_1(f)|$ (see Figure 3.22b). Since jitter peaking was present, an exponential rms jitter accumulation was expected. Figure 3.23b shows the output jitter transient photograph. This output transient shows exponential jitter accumulation as expected. The transient was divided into 330 μs sections, and $\sigma^L[N]$ for successive passes of data around the loop was calculated. In Figure 3.24, $\sigma^L[N]$ is plotted versus N along with the model calculations given by (3.9) using a jitter transfer function with 0.15 dB jitter peaking. Again, in both the calculation and the simulation, exponential growth was predicted and measured up to 125 cascade regenera-

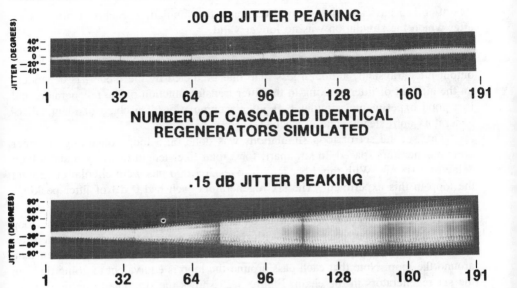

Fig. 3.23 Output jitter transient for 191 passes around a loop containing one regenerator and 69 km of fiber

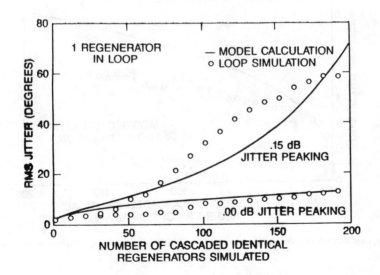

Fig. 3.24 Comparison of loop simulation results with the accepted model of jitter accumulation

tors after 125 passes of data around the loop, the simulation reached a steady state after which the simulation was no longer valid.

The results of the circulating loop agree qualitatively with the model predictions when jitter peaking is present and not present. These results show that the loop's output jitter transient, although less than that predicted by the model, is dependent on the amount of jitter peaking in the jitter transfer function of the regenerator, and is a good experimental simulator of jitter accumulation of a chain of identical regenerators until steady state is reached.

Our second set of loop simulations was done on a loop containing six fiber optic regenerators spaced 36 km apart, for a total fiber length of 217 km and a loop delay of 1.05 ms. With this fiber length, over 307,000 bits were circulating around the loop in this experiment. The six regenerators each had 0 dB of jitter peaking, so roughly $\sqrt{N}$ jitter accumulation was expected.

The loop delay was trimmed to exactly an integer number of bits, the loop closed, and the output jitter of the sixth regenerator recorded. By dividing the jitter transient into 1.05 ms sections, we calculated $\sigma^L[N]$ for successive passes of data around the loop. Note that each pass around the loop is equivalent to going through the six regenerators in the chain. Figure 3.25 shows the simulated rms jitter accumulation versus the number of regenerators along the jitter model calculation. Examining Figure 3.25 we see that the simulated jitter accumulation was close to that calculated for up to 200 cascaded regenerators.

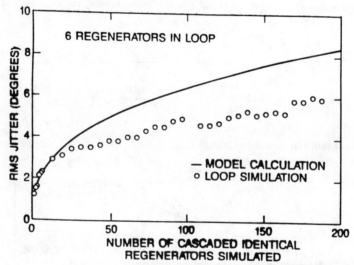

Fig. 3.25 Comparison of loop simulation results with the accepted model of jitter accumulation

Our last loop simulation closed a loop on 21 regenerators spaced an average of 35.4 km apart and simulated the jitter accumulation of a longer chain of regenerators. With this many regenerators, we had a total fiber length of 744 km and a loop delay of 3.6 ms. Therefore, in this simulation over one million bits were circulating around the loop. These 21 regenerators each had 0.00 dB measured jitter peaking, so a nonexponential jitter accumulation was expected.

The loop delay was trimmed to exactly an integer number of bits, the loop closed, and the output jitter transient of the twenty-first generator recorded. Figure 3.26 shows the output jitter transient waveform. By dividing this recording into 3.6 ms sections, $\sigma^L[N]$ for successive passes of data around the loop was calculated. Note that each pass around the loop is equivalent to going through all 21 regenerators in the chain. Figure 3.27 shows $\sigma^L[N]$ versus N, along with the jitter model calculation. Figure 3.27 shows that the jitter accumulation was less than $\sqrt{N}$, and agreed well with the model's prediction. A reason for the closer agreement is that over one million bits were circulating and only 11 passes around the loop were needed to simulate 231 cascaded regenerators.

We have shown results of several jitter accumulation measurements that were obtained by circulating loop simulation. We were able to obtain results that qualitatively agreed with the jitter accumulation models presented in Section 3.2. Although the circulating loop has a limit to the number of passes that will accurately simulate the jitter accumulation of a chain of identical regenerators [3.19], we have shown that the loop simulation has value as an experimental technique for measuring and characterizing jitter accumulation of fiber optic regenerators. In particular, the loop simulation technique allows a system designer to experimentally compare the system jitter accumulation characteristics of one regenerator design against another regenerator design without the need of building many test units. Therefore the loop simulation technique is a valuable tool for characterizing jitter accumulation in digital equipment.

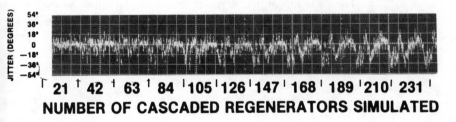

NUMBER OF CASCADED REGENERATORS SIMULATED

Fig. 3.26 Output jitter transient for 11 passes around a loop containing 21 regenerators and 744 km of fiber

3.4 A STOCHASTIC JITTER ACCUMULATION MODEL

In Section 3.1 we described the jitter accumulation model. For this model the jitter transfer function and jitter power spectral density of each regenerator are different and need to be known. This model would be useful only when steps have been taken to measure each regenerator individually. If a system is being designed, has already

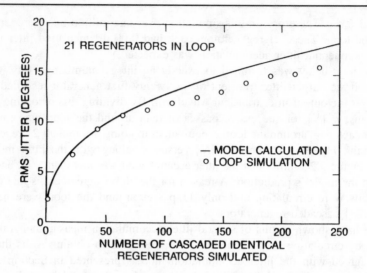

Fig. 3.27 Comparison of loop simulation results with the accepted model of jitter accumulation

been deployed, or when analyzing the results of the circulating loop experiment, the specialized identical regenerator model is useful. A question frequently asked about the identical model is how variations in the jitter transfer function of the regenerators affect the jitter accumulation. To study this we look at a stochastic jitter accumulation model given by Chamzas [3.26]. The stochastic model considers the jitter transfer function in each regenerator to be an independent and identically distributed random variable and calculates the mean and variance of the accumulated rms systematic and random jitter. The variation in jitter transfer functions can either be the design tolerance allowed in the manufacture of the timing circuits or it may account for the aging of the components with time.

In either case, this model assumes that the jitter spectral densities generated by each regenerator are known, identical, and statistically white, so that

$$\Phi_{il}^R = \Phi_{i1}^R \quad \text{and} \quad \Phi_{il}^S = \Phi_{i1}^S \quad \text{for all} \quad l \in [1,N] \tag{3.28}$$

This simplifies (3.3) and (3.6), respectively, to

$$\Phi_N^R(f) = \Phi_{i1}^R \left[\prod_{l=1}^{N} |H_l(f)|^2 + \prod_{l=2}^{N} |H_l(f)|^2 + \ldots + |H_N(f)|^2 \right] \tag{3.29}$$

and

$$\Phi_N^S(f) = \Phi_{i1}^S \left[\left| \prod_{l=1}^{N} H_l(f) + \prod_{l=2}^{N} H_l(f) + \ldots + H_N(f) \right|^2 \right] \tag{3.30}$$

The model then assumes that $H_l(f)$ of the regenerators are independent and identically distributed so that their first and second moments are equal. We then find the mean and variance of $\Phi_N^R(f)$, as [3.26]

$$E\{\Phi_N^R(f)\} = \Phi_{i1}^R E\{|H_l(f)|^2\} \frac{1 - E\{|H_l(f)|^2\}^N}{1 - E\{|H_l(f)|^2\}} \tag{3.31}$$

and

$$\mathrm{Var}\{\Phi_N^R(f)\} = \sqrt{E\{\Phi_N^R(f)^2\} - E^2\{\Phi_N^R(f)\}} \tag{3.32}$$

where

$$E\{\Phi_N^R(f)^2\} = [\Phi_{i1}^R]^2 \left[E\{|H_l(f)|^4\} \frac{1 - E\{|H_l(f)|^4\}^N}{1 - E\{|H_l(f)|^4\}} \right.$$

$$+ 2 \frac{E\{|H_l(f)|^4\}}{1 - \dfrac{E\{|H_l(f)|^4\}}{E\{|H_l(f)|^2\}}}$$

$$\left. \cdot \left\{ \frac{1 - E\{|H_l(f)|^2\}^N}{1 - E\{|H_l(f)|^2\}} - \frac{1 - E\{|H_l(f)|^4\}^N}{1 - E\{|H_l(f)|^4\}} \right\} \right] \tag{3.33}$$

We then find the mean and variance of $\Phi_N^S(f)$, as

$$E\{\Phi_N^S(f)\} = \Phi_{i1}^S \left[E\{|H_l(f)|^2\} \frac{1 - E\{|H_l(f)|^2\}^N}{1 - E\{|H_l(f)|^2\}} \right.$$

$$+ 2 \, \mathrm{Re} \frac{E\{|H_l(f)|^2\}}{1 - \dfrac{E\{|H_l(f)|^2\}}{E\{H_l(f)\}}}$$

$$\left. \cdot \left\{ \frac{1 - E\{H_l(f)\}^N}{1 - E\{H_l(f)\}} - \frac{1 - E\{|H_l(f)|^2\}^N}{1 - E\{H_l(f)\}} \right\} \right] \tag{3.34}$$

and

$$\text{Var}\{\Phi_N^S(f)\} = (\Phi_{i1}^S)\left[E\{|H_l(f)|^2\}\frac{1 - E^N\{|H_l(f)|^2\}}{1 - E\{|H_l(f)|^2\}}\right.$$

$$+ 2\text{Re}\frac{E\{|H_l(f)|^2\}}{1 - \dfrac{E\{|H_l(f)|^2\}}{E\{H_l(f)\}}}$$

$$\cdot\left\{\frac{1 - E^N\{H_l(f)\}}{1 - E\{H_l(f)\}} - \frac{1 - E^N\{|H_l(f)|\}}{1 - E\{|H_l(f)|\}}\right]$$

$$+ [2Im + \text{Re}^2]\left[E\{H_l^2(f)\}\frac{1 - E^N\{H_l^2(f)\}}{1 - E\{H_l^2(f)\}}\right.$$

$$+ 2\frac{E\{H_l^2(f)\}}{1 - \dfrac{E\{H_l^2(f)\}}{E\{H_l(f)\}}}\left[\frac{1 - E^N\{H_l(f)\}}{1 - E\{H_l(f)\}}\right.$$

$$\left.\left.\left.- \frac{1 - E^N\{H_l^2(f)\}}{1 - E\{H_l(f)\}}\right]\right]^2\right]$$

$$- 2\left|E\{H_l(f)\}\frac{1 - E^N\{H_l(f)\}}{1 - E\{H_l(f)\}}\right|^2 \tag{3.35}$$

With the mean and variance of the random and systematic jitter spectrums we can find the expected value of the rms jitter accumulation as

$$E\{\sigma_R[N]\} = \sqrt{\int_{-1/T}^{1/T} E\{\Phi_N^R(f)\}df} \tag{3.36}$$

and

$$E\{\sigma_S[N]\} = \sqrt{\int_{-1/T}^{1/T} E\{\Phi_N^S(f)\}df} \tag{3.37}$$

We can establish the 99% confidence bounds on the rms jitter by calculating

$$\sigma_R^{99\%}[N] = \sqrt{\int_{-1/T}^{1/T} [E\{\Phi_N^R(f)\} \pm 3\mathrm{Var}\{\Phi_N^R(f)\}]df} \qquad (3.38)$$

$$\sigma_S^{99\%}[N] = \sqrt{\int_{-1/T}^{1/T} [E\{\Phi_N^S(f)\} \pm 3\mathrm{Var}\{\Phi_N^S(f)\}]df} \qquad (3.39)$$

To illustrate this stochastic jitter accumulation model in an example, we assume that for each fixed f, $H_l(f)$ for $l = 1$ to N is a uniformly distributed random variable between $H_a(f)$ and $H_b(f)$. $H_a(f)$ is a second-order low-pass filter given by (3.16) with $\xi = 1/\sqrt{2}$ (no jitter peaking) and $H_b(f)$ is given by (3.16) with $\xi = 0.625$ (0.2 dB jitter peaking). Therefore, we are randomly selecting $H_l(f)$ of the regenerators in the chain with magnitudes in the region between curves a and d of Figure 3.13. Thus any regenerator can have jitter peaking ranging from 0.00 dB to 0.20 dB.

For a uniform distribution of $H_l(f)$, the following moments are calculated

$$E\{H_l(f)\} = \frac{H_a(f) + H_b(f)}{2} \qquad (3.40)$$

$$E\{H_l^2(f)\} = \frac{H_a^2(f) + H_a(f)\,H_b(f) + H_b^2(f)}{3} \qquad (3.41)$$

$$E\{|H_l(f)|^2\} = \frac{|H_b(f)|^2 + |H_a(f)|^2}{2} \qquad (3.42)$$

$$E\{|H_l(f)|^4\} = \frac{|H_b(f)|^4 + |H_a(f)|^2\,|H_b(f)|^2 + |H_b(f)|^4}{3} \qquad (3.43)$$

Inserting (3.42) and (3.43) in (3.36), we plot in Figure 3.28 $E\{\sigma_R[N]\}$ along with the upper and lower $\sigma_R^{99\%}[N]$. Inserting (3.40) and (3.41) in (3.37), we plot $E\{\sigma_S[N]\}$ in Figure 3.29 along with upper and lower 99% confidence bounds. Examining Figures 3.28 and 3.29, and comparing them with the identical regenerator model shown in Figures 3.15 and 3.16, we can conclude that the identical regenerator jitter accumulation model will give a reliable estimate of the rms jitter accumulation if the average jitter transfer function is used as a parameter.

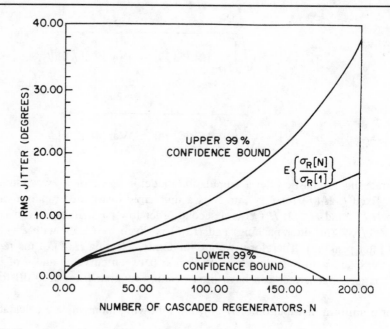

Fig. 3.28 Normalized random rms jitter accumulation for statistically distributed jitter transfer functions

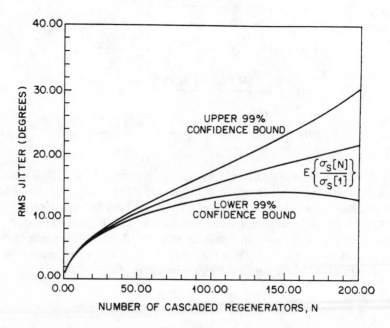

Fig. 3.29 Normalized systematic rms jitter accumulation for statistically distributed jitter transfer functions

3.5 REMARKS

In this chapter, we have studied the accumulation of jitter introduced by line regenerators, both analytically and experimentally. Analytically, we modeled jitter accumulation assuming nonidentical, identical, and statistically distributed regenerators. Experimental results from jitter accumulation measurements and circulating loop simulations showed the validity of the jitter accumulation models.

REFERENCES

[3.1]   P. R. Trischitta and P. Sannuti, "Accumulation of Pattern-Dependent Jitter for a Chain of Fiber Optic Regenerators," *IEEE Transactions on Communications*, Vol. COM-36, No. 6, June 1988.

[3.2]   D. G. Ross, R. M. Paski, D. G. Ehrenberg, and G. M. Homsey, "A Highly Integrated Regenerator for 295.6 Mb/s Undersea Optical Transmission," *IEEE/OSA Journal of Lightwave Technology*, Vol. LT-2, No. 6, December 1984.

[3.3]   C. J. Byrne, B. J. Karafin and D. B. Robinson, Jr., "Systematic Jitter in a Chain of Digital Regenerators," *Bell System Technical Journal*, November 1963.

[3.4]   J. T. Harvey and J. W. Rice, "Random Timing Noise Growth in a Cascaded Digital Regenerator Chain," *IEEE Transactions on Communications*, August 1973.

[3.5]   H. Yamamoto and S. Kuba, "Performance of Timing Recovery Circuits with Phase-Locked Loop in a Long Chain of Regenerative Repeaters," *Electronics and Communications in Japan*, Vol. 59-B, No. 3, 1976.

[3.6]   T. Shimamura and I. Eguchi, "Analysis of Jitter Accumulation in a Chain of PLL Timing Recovery Circuits," *IEEE Transactions on Communications*, Vol. COM-25, No. 9, September 1977.

[3.7]   E. L. Varma and J. Wu, "Analysis of Jitter Accumulation in a Chain of Digital Regenerators," *Proceeding of the IEEE Globecom*, 1982, Vol. II, pp. 653–657.

[3.8]   E. O. Sunde, "Self-Timing Regenerative Repeaters," *Bell System Technical Journal*, pp. 891–937, July 1957.

[3.9]   E. Roza, "Analysis of Phase-Locked Timing Extraction Circuits for Pulse Code Transmission," *IEEE Transactions on Communications*, Vol. COM-22, No. 9, September 1979.

[3.10] D. L. Dittweiler, "The Jitter Performance of Phase-Locked Loops Extracting Timing From Baseband Data Waveforms," *Bell System Technical Journal*, Vol. 55, No. 1, January 1976.

[3.11] D. A. Fishman, R. L. Rosenberg, C. Chamzas, "Analysis of Jitter Peaking Effects in Digital Long-Haul Transmission Systems Using SAW-Filter Retiming," *IEEE Transactions on Communications*, July 1985.

[3.12] U. Mengali and G. Pirani, "Jitter Accumulation in PAM Systems," *IEEE Transactions on Communications*, Vol. COM-28, No. 8, August 1980.

[3.13] L. E. Franks and J. P. Bubrouski, "Statistical Properties of Timing Jitter in a PAM Timing Recovery Scheme," *IEEE Transactions on Communications*, Vol. COM-22, No. 7, July 1974.

[3.14] J. M. Manley, "The Generation and Accumulation of Timing Noise in PCM Systems," *Bell System Technical Journal*, Vol. 48, March 1969.

[3.15] L. R. Wrathall, "Transistorized Binary Pulse Regenerator," *Bell System Technical Journal*, Vol. 35, No. 5, September 1956.

[3.16] R. L. Carbrey and C. J. Byrne, "A Long Delay Circulating Loop Test for PCM Timing and Repeater Studies," Unpublished Report, AT&T Bell Laboratories, March 9, 1962.

[3.17] B. K. Kinariwala, "Timing Jitter in a Circulating Loop," *Symposium on Signal Transmission and Processing*, Columbia University 1965, Conference Record, p. 117–120.

[3.18] P. R. Trischitta, P. Sannuti and C. Chamzas, "A Circulating Loop Experimental Technique to Simulate the Jitter Accumulation of a Chain of Fiber Optic Regenerators," *IEEE Transaction on Communications*, Vol. COM-36, No. 2, February 1988.

[3.19] P., G. Hoel, *Elementary Statistics*, John Wiley & Sons, 1960.

[3.20] C. Chamzas and P. R. Trischitta, "Simulation of a Chain of Digital Optoelectronic Regenerators," *Proceedings of the International Modeling and Simulation Conference*, Paper 3.4-13, Athens, Greece, June 27–29, 1984.

[3.21] S. M. Abbott, R. E. Wagner, P. R. Trischitta, "SL Undersea Lightwave System Experiments," *Proceedings of ICC '83*, Paper C.5, Boston, Massachusetts, June 19–23, 1983.

[3.22] R. E. Wagner, S. M. Abbott, P. R. Trischitta, "Undersea Lightwave Transmission Experiments," *Proceedings of the Fourth International Conference on Integrated Optics and Optical Fiber Communications*, Paper 29A41, Tokyo, Japan, June 1983.

[3.23] J. H. Wilbrod and J. Moulu, "Jitter Accumulation in Optical Transmission Systems," *Electronics Letters*, Vol. 21, No. 17, August 15, 1985.

[3.24] M. Amemiya, Y. Hayashi and T. Ito, "Systematic Jitter Suppression by Pattern Inversion," *Electronics Letters*, June 26, 1985.

[3.25] S. R. Nagel, "Review of the Depressed Cladding Single-Mode Fiber Design and Performance for the SL Undersea System Application," *IEEE/OSA Journal of Lightwave Technology*, Vol. LT-2, No. 6, December 1984.

[3.26] C. Chamzas, "Accumulation of Jitter: A Stochastic Model," *AT&T Technical Journal*, Vol. 64, No. 1, January 1985.

Chapter 4
The Effect of Jitter on Transmission Quality

Thus far in this book we have studied how jitter is generated in line regenerators and how this jitter accumulates when regenerators were cascaded. However, we have yet to mention the effect accumulated jitter has on transmission quality, and in particular, how accumulated jitter effects the probability of making correct bit decisions inside the regenerator. In this chapter, we will analyze and describe ways to minimize the effect accumulated jitter has on the decision-making process inside the Nth regenerator of a long chain of cascaded regenerators.

By defining alignment jitter as the difference between the jitter on the received signal to be sampled and the jitter on the extracted clock doing the sampling, we will show that it is accumulated alignment jitter that will degrade the regenerator's performance. We will then derive the criteria for a correct bit decision in the presence of receiver noise, static phase offset, and accumulated alignment jitter, and calculate the transmission penalty caused by specific distributions of accumulated alignment jitter. To assure satisfactory performance of manufactured regenerators that are to operate in multiregenerator systems, we will define the jitter tolerance of a regenerator as an effective measure of a regenerator's capability to tolerate incoming accumulated jitter. We will show that excessive transmission penalty will not occur if the accumulated jitter does not exceed the regenerator's measured jitter tolerance template. Finally, we will present models for alignment jitter accumulation to compare this chapter with the jitter accumulation models of Chapter 3.

4.1 CORRECT BIT DECISIONS IN THE PRESENCE OF JITTER

In Figure 4.1, we show again the block diagram of a self-timed regenerator. Recall the signal received by the Nth regenerator in a long chain of regenerators is corrupted by receiver noise and accumulated jitter from the previous $N - 1$ regenerators, and is given by

$$r_N(t) = \sum_{n=-\infty}^{\infty} a_n\, g_N(t - nT - e_{N-1}[nT]) + \eta_N(t) \qquad (4.1)$$

where $a_n \in \{0,1\}$ and $g_N(t)$ is the amplified and equalized pulse shape. This received signal is sampled by the decision circuit and a regenerated output signal $r'(t)$ is produced based on decisions made about $r_N(t)$ at the sampling instances. The signal for sampling comes from the timing extraction circuit and is given by

$$s_N(t) = |A_N| \sin\left[\frac{2\pi}{T}(t - e_N(t) - \tau)\right] \qquad (4.2)$$

at the decision circuit, where τ is a static delay because of the physical separation of the timing and decision circuits. When $s_N(t)$ rises to a level V, $r_N(t)$ is sampled, and the decision circuit outputs a regenerated signal that in turn modulates a semiconductor laser transmitter. The regenerated output signal is expressed as

$$r'_N(t) = \sum_{n=-\infty}^{\infty} a'_n f_N(t - nT - e_N[nT] - \tau - t_0) \qquad (4.3)$$

where a'_n is the regenerated bit sequence, $f_N(t)$ is the output pulse shape, $t_0 = T/2\pi$ $\sin^{-1} V/|A_N|$ is a static delay caused by the sampling threshold V, and $e_N[nT]$ is the output jitter sequence. The regenerated bit sequence of (4.3) is given by

$$a'_n = \text{step}[r_N(t_n) - E\{a_n\}g_N(0)] \qquad (4.4)$$

where $E\{a_n\}g_N(0)$ is the decision threshold, and t_n represents the sampling times given by

$$t_n = t_0 + \tau + e_N[nT] + nT \qquad (4.5)$$

Examining (4.5), we see that the sampling times t_n are shifted from exact multiples of nT by two static terms and one jitter term [4.1]. The static t_0 term is caused by a non-zero sampling threshold of the decision circuit. A $V = 0$ sampling threshold eliminates this static phase shift; therefore, DC offsets in the sampling portion of the decision circuit should be minimized. The static phase offset τ is caused by the physical separation of the timing and decision circuits. This static phase alignment between the data and the timing signals must be set either by choosing the length of a coaxial cable between the timing and decision circuits or by some electronic phase shifting network so that the data signal is sampled at an optimum time.

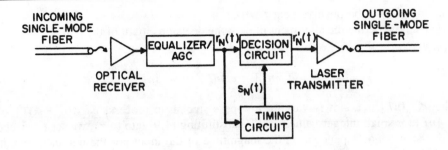

Fig. 4.1 Block diagram of self-timed regenerator

Substituting (4.5) and (4.1) into (4.4), and noting that $E\{a_n\} = 1/2$ for $a_n \in \{0,1\}$, the regenerated bit sequence is

$$a_n' = \text{step}\left[\sum_{m=-\infty}^{\infty} a_m g_N(nT - mT + t_o + \tau + e_N[nT]\right.$$

$$\left. - e_{N-1}[mT]) + \eta_N(t_n) - \frac{1}{2} g_N(0)\right] \tag{4.6}$$

Changing indices to $k = n - m$ and extracting the $k = 0$ term yields

$$a_n' = \text{step}\left[a_n g_N(t_0 + \tau + e_N[nT] - e_{N-1}[nT])\right.$$

$$+ \sum_{\substack{k=-\infty \\ k \neq 0}}^{\infty} a_{n-k} g_N(kT + t_0 + \tau + e_N[nT] - e_{N-1}[(n - k)T])$$

$$\left. + \eta_N(t_n) - \frac{1}{2} g_N(0)\right] \tag{4.7}$$

Examining (4.7), we see that a correct decision will be made, i.e., $a_n' = a_n$ if

$$\frac{1}{2} g_N(0) - g_N(t_o + \tau + e_N[nT] - e_{N-1}[nT]) - \sum_{k \neq 0} a_{n-k} g_N(kT + t_o$$

$$+ \tau + e_N[nT] - e_{N-1}[(n - k)T] < \eta_N(t_n) \quad \text{if} \quad a_n = 1 \tag{4.8a}$$

or

$$\frac{1}{2} g_N(0) - \sum_{k \neq 0} a_{n-k} g_N(kT + t_o + \tau + e_N[nT] - e_{N-1}[(n - k)T])$$

$$> \eta_N(t_n), \quad \text{if} \quad a_n = 0 \tag{4.8b}$$

assuming $g_N(t)$ to be positive near $t = 0$.

We now define the alignment jitter at the decision circuit of the Nth regenerator as

$$e_{aN}[nT] = e_N[nT] - e_{N-1}[nT] \qquad (4.9)$$

Since $e_{N-1}[nT]$ is a slowly-varying low-pass stochastic process, $e_{N-1}[(n - k)T] \approx e_{N-1}[nT]$ for small integer values of $|k|$. Substituting (4.9) into (4.8), we see a correct bit decision is not dependent on the magnitude of the input nor the output jitter but the difference between the two (i.e., the alignment jitter). Therefore it is $e_{aN}[nT]$ that will degrade the regenerator's ability to make correct decisions, not $e_N[nT]$ nor $e_{N-1}[nT]$. Rewriting (4.8) using (4.9) and noting that the effect of a non-zero decision threshold is to add another static time shift to the sampling time, we combine t_0 and τ into one static phase offset term represented by τ and write (4.8) as

$$\frac{1}{2} g_N(0) - g_N(e_{aN}[nT] + \tau) - \sum_{\substack{k=-\infty \\ k \neq 0}}^{\infty} a_{n-k}\, g_N(kT + e_{aN}[nT] + \tau) < \eta_N(t_n)$$

$$\text{if } a_n = 1 \qquad (4.10a)$$

or

$$\frac{1}{2} g_N(0) - \sum_{\substack{k=-\infty \\ k \neq 0}}^{\infty} a_{n-k}\, g_N(kT + e_{aN}[nT] + \tau) > \eta_N(t_n) \quad \text{if} \quad a_n = 0 \qquad (4.10b)$$

Equation (4.10) is the criterion for a correct bit decision in the presence of noise, alignment jitter, and a static phase offset. We see that making a correct bit decision is dependent on the received pulse shape, the decision threshold, the bit sequence before and after the present bit decision, the alignment jitter, the static phase offset and, of course, noise.

To calculate the bit error probability P_E, we must assume that $g_N(t)$ is time-limited, and only K past pulses and K future pulses significantly affect the decision-making process. Thus only $\pm K$ bits need be considered in the infinite series of (4.10). For a particular bit i in a $2K + 1$ bit message sequence, where zeroes and ones are equally likely, the probability of making a bit error is

$$P_{Ei}(e_{aN}[nT] + \tau) = \frac{1}{2} P\Bigg(g_N(e_{aN}[nT] + \tau)$$

$$+ \sum_{\substack{k=-K \\ k \neq 0}}^{K} a_{n-k}\, g_N(kT + e_{aN}[nT] + \tau)$$

$$-\frac{1}{2} g_N(0) < \eta_N(t_n) \Big) + \frac{1}{2} P\Big(\frac{1}{2} g_N(0)$$

$$-\sum_{\substack{k=-K \\ k \neq 0}}^{K} a_{n-k}\, g_N(kT + e_{aN}[nT] + \tau) < \eta_N(t_n)\Big) \quad (4.11)$$

where $P(x)$ is the probability of x. To proceed, we must consider the statistics of the noise process $\eta_N(t)$. In fiber optic transmission systems, $\eta_N(t)$ is made up of receiver shot noise caused by the statistical nature of the photon-electron interaction and receiver thermal noise from the receiver amplifier electronics. Shot noise is Poisson distributed and signal dependent (i.e., present when an optical pulse is received and absent when no pulse is received). Thermal noise, on the other hand, is signal-independent and Gaussian distributed. If the photodiode shot noise is the same order of magnitude as the receiver's amplifier noise, the calculation of P_E for a given alignment jitter is complex [4.2]. However, if shot noise can be neglected compared with thermal noise, the calculation of P_E, for a fiber optic system is the same as in classical digital systems.

For fiber optic systems using direct detection and PIN photodiodes or APDs with average gains less than twenty, shot noise can be neglected. In this book we are most interested in analyzing the effect of jitter on P_E. In addition, our experimental work was done using PIN photodiode receivers; therefore we will neglect shot noise in our analysis. Here we assume that the sampled noise process $\eta_N(t_n)$ is Gaussian distributed with a flat power spectral density of N_0 Watts/Hz resulting in [4.3–4.5]

$$P_{Ei}(e_{aN}[nT] + \tau) = \frac{1}{4}\,\text{erfc}$$

$$\cdot \left(\frac{g_N(e_{aN}[nT] + \tau) + \displaystyle\sum_{\substack{k=-K \\ k \neq 0}}^{K} a_{n-k} g_N(kT + e_{aN}[nT] + \tau) - \frac{1}{2} g_N(0)}{\sqrt{N_0}} \right)$$

$$+ \frac{1}{4}\,\text{erfc}\left(\frac{\dfrac{1}{2} g_N(0) - \displaystyle\sum_{\substack{k=-K \\ k \neq 0}}^{K} a_{n-k} g_N(kT + e_{aN}[nT] + \tau)}{\sqrt{N_0}} \right) \quad (4.12)$$

where

$$\text{erfc}(x) = \frac{2}{\sqrt{\pi}} \int_x^\infty e^{-y^2} dy$$

If we further assume that all the 2^{2K} possible $2K + 1$ bit sequences are equally likely, then P_E for a given amount of alignment jitter and static phase offset is

$$P_E(e_{aN}[nT] + \tau) = \frac{1}{2^{2K}} \sum_{i=1}^{2^{2K}} P_{Ei}(e_{aN}[nT] + \tau) \qquad (4.13)$$

Since $e_{aN}[nT]$ is a stochastic process, the average P_E is

$$\bar{P}_E = \int_{-\infty}^{\infty} P_E(e_{aN}[nT] + \tau) p(e_{aN}[nT]) de_{aN}[nT] \qquad (4.14)$$

where $p(e_{aN}[nT])$ is the probability density function of $e_{aN}[nT]$.

4.2 JITTER RELATED TRANSMISSION PENALTY

To calculate the *transmission* penalties caused by a specific distribution of alignment jitter, we will assume an ideal raised cosine pulse shape [4.6] given by

$$g_N(t) = \frac{\sin \pi t/T}{\pi t/T} \frac{\cos \pi t/T}{1 - 4t^2/T^2} \qquad (4.15)$$

and $K = 7$ adjacent pulses affecting the current pulse decision.

We can examine the penalty caused by static phase offset alone by assuming $e_{aN}[nT]$ is zero and calculating the P_E for various τ where we converted τ to units of degrees by defining T seconds $= 360°$. Figure 4.2 shows the P_E given by (4.13) as a function of the *signal to noise ratio* (SNR) for various τ. The SNR is given by the argument of the complimentary error function in (4.12). Figure 4.2 shows that a transmission penalty results from not sampling at the instant when the pulse is at its peak value. Transmission penalty is defined as the extra SNR for a given P_E, when compared with the zero jitter and static phase offset case. Given that 1×10^{-9} is the required P_E we are considering, we graph in Figure 4.3 the additional SNR required to maintain a P_E of 1×10^{-9} (i.e., the bit error rate (BER) penalty) versus τ, which also shows that the penalty accelerates as the static phase offset increases, following a parabolic shape for the raised cosine pulse shape. Also from Figure 4.3 we note that with zero accumulated alignment jitter, the penalty is less than 0.5 dB for static phase offsets as large as $34°$ from optimum. Thus, with no alignment jitter, the effect of optimum static phase on transmission performance is small.

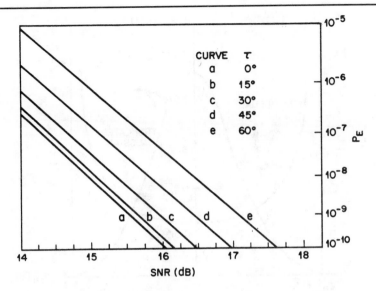

Fig. 4.2 The effect static phase offset τ has on P_E

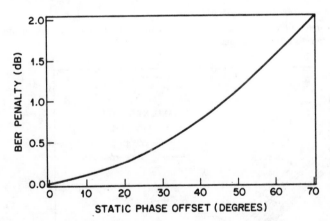

Fig. 4.3 Transmission penalty at $P_E = 1 \times 10^{-9}$ caused by static phase offset

This is not the case when significant accumulated alignment jitter is received by the regenerator. The static phase offset is then important, and must be minimized. We can show this by assuming $e_{aN}[nT]$ has a truncated Gaussian distribution, with a zero mean and a standard deviation $e_{aN}^{P-P}/6$. The distribution is truncated at half the maximum allowable peak-to-peak alignment jitter of e_{aN}^{P-P}. This distribution, shown in Figure 4.4, is the most likely distribution of the accumulated jitter from many cascaded fiber optic regenerators. Using (4.14) we graph in Figure 4.5 the P_E for peak-to-peak alignment jitter of $0°$, $50°$, and $100°$ with static phase offsets of $0°$, $15°$ and $30°$. It is readily seen that the transmission penalty is largest when a combination

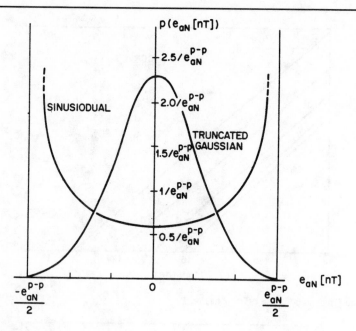

Fig. 4.4 Comparison of the probability density functions of sinusoidal alignment jitter and truncated Gaussian distributed alignment jitter

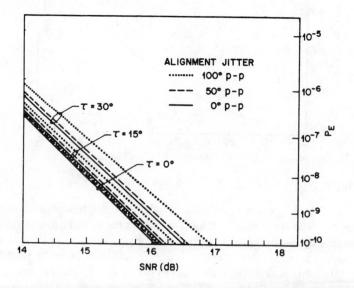

Fig. 4.5 The combined effect of static phase offset τ and truncated Gaussian distributed alignment jitter on P_E

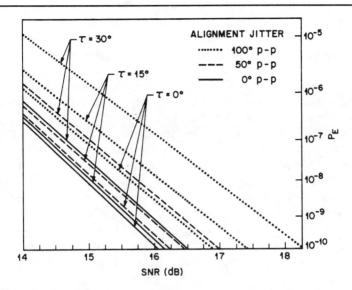

Fig. 4.6 The combined effect of static phase offset τ and sinusoidal alignment jitter on P_E

of alignment jitter and static phase offset is present at the Nth regenerator's decision circuit. This shows that the regenerator is less tolerant to accumulated alignment jitter when static phase offset is present. Therefore, when designing and manufacturing regenerators, static phase offset must be minimized in order to minimize the effect accumulated alignment jitter has on the P_E.

As a final calculation using (4.14), we assume that $e_{aN}[nT]$ is a sinusoid. Characterizing the effect of sinusoidal alignment jitter on P_E will be important in the next section on jitter tolerance. The probability density function for sinusoidal alignment jitter [4.7] is shown in Figure 4.4, where it can be compared with the truncated Gaussian distribution. Again using (4.14), we graph the P_E for peak-to-peak sinusoidal alignment jitter of $0°$, $50°$ and $100°$ and static phase offsets of $0°$, $15°$ and $30°$. We again see that the transmission penalty is largest when a combination of alignment jitter and static phase offset is present. In Figure 4.7, we have graphed the penalty at $P_E = 1 \times 10^{-9}$ for the sinusoidal and the truncated Gaussian distributed alignment jitter cases. Figure 4.7 shows that in general, with and without static phase offsets, sinusoidal alignment jitter, with a peak-to-peak value of e_{aN}^{p-p}, results in penalties that are about one dB larger than those caused by truncated Gaussian distributed alignment jitter of the same peak-to-peak value.

4.3 REGENERATOR JITTER TOLERANCE

In the previous sections, we analyzed the effect that imperfect sampling has on the

P_E of a regenerator. We showed in terms of BER penalties at $P_E = 1 \times 10^{-9}$ how alignment jitter, static phase offset, and receiver noise degrade regenerator performance. But using (4.14) to assure satisfactory performance of manufactured regenerators that are destined to operate somewhere in a long chain of regenerators is extremely difficult. For manufactured regenerators, the parameters that must be inserted into (4.14) are either difficult to measure accurately, or are unknown. Specifically, at the Nth regenerator the received pulse shape $g_N(t)$ is dependent on the previous laser's output optical pulse shape, and on any chromatic dispersion in the single-mode fiber span, so it is difficult to measure effectively during regenerator manufacture. The accumulated alignment jitter is of unknown distribution and magnitude and τ is difficult to set and may be temperature dependent. Also, the noise sensitivities of the optical receivers in each regenerator are not easily characterized. Therefore, difficulties exist in directly applying (4.14) to assess the system performance of manufactured fiber optic regenerators. To assure satisfactory system performance of fiber optic regenerators before system installation, we define a jitter tolerance measurement, that although coupled to (4.14), overcomes some of its limitations.

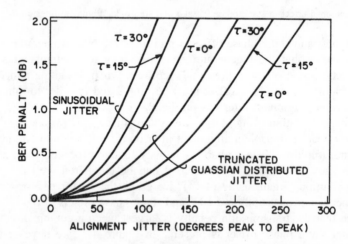

Fig. 4.7 The BER penalty at 1×10^{-9} caused by known distributions of accumulated alignment jitter into a regenerator with a static phase offset τ

Jitter tolerance is a measure of the capability of a regenerator to tolerate incoming jitter [4.8–4.10]. We can formally define the jitter tolerance of a regenerator as the magnitude of incoming jitter that results in a specific transmission penalty. Jitter tolerance is of considerable importance when manufacturing or designing regenerators for use in long-haul transmission systems. It can be used to assure that an excessive penalty will not occur when the regenerator is placed into a long system that has an input jitter from the previous regenerators not exceeding the regenerator's measured jitter tolerance curve. In this way the regenerator's performance during design and manufacture can be coupled to optimizing overall system performance.

In the previous section, we showed that sinusoidal jitter has a distribution that causes a much larger penalty than jitter with a truncated Gaussian distribution. Therefore, we use sinusoidal input jitter in defining the jitter tolerance of a regenerator, since it can be considered as having a worse case distribution for most digital transmission systems. The input jitter to the regenerator under test is then constructed to be

$$e_{N-1}[nT] = K_i \sin 2\pi f_{pm}t\big|_{t=nT} \tag{4.16}$$

where $2K_i$ is the peak-to-peak input jitter and f_{pm} is the jitter frequency. Using a large sinusoidal jitter ($e_{N-1}[nT] \gg e_{iN}[nT]$) as input to the regenerator results in an output jitter of

$$e_N[nT] = K_i |H_N(f_{pm})| \sin (2\pi f_{pm}t + \arg H_N(f_{pm}))\big|_{t=nT} \tag{4.17}$$

where $H_N(f_{pm})$ is the jitter transfer function of the regenerator under test. Subtracting (4.16) from (4.17) results in an alignment jitter of

$$e_{aN}[nT] = K_i(1 - H_N(f_{pm})) \sin 2\pi f_{pm}nT \tag{4.18}$$

where the peak-to-peak alignment jitter is

$$e_{aN}^{P-P} = 2K_i|1 - H_N(f_{pm})| \tag{4.19}$$

Equation (4.19) shows that the peak-to-peak alignment jitter is dependent on the frequency of the input jitter. At low frequencies, where $|H_N(f_{pm})| \approx 1$, (4.19) shows that large peak-to-peak input jitter results in small peak-to-peak alignment jitter. At high frequencies, where $|H_N(f_{pm})| \approx 0$, the peak-to-peak alignment jitter equals the peak-to-peak input jitter. In Section 4.1, the penalty was shown to be dependent on peak-to-peak accumulated alignment jitter. Therefore, (4.19) shows that the regenerator can tolerate large input jitter at low frequencies, but cannot tolerate much input jitter at high frequencies. By selecting a BER penalty that is small and tolerable (e.g., .5 dB), the peak-to-peak input jitter can be measured as a function of the jitter frequency f_{pm} which causes this tolerated BER penalty. This measurement will trace out a jitter tolerance template that, if not exceeded, will assure that the regenerator performs with less than the tolerated BER penalty. This is because sinusoidal jitter stresses a regenerator much more so than the expected Gaussian distribution of accumulated jitter. Thus by measuring the jitter tolerance of a regenerator, and requiring the regenerator meet a particular jitter tolerance requirement, the specified system performance of a particular regenerator can be assured during manufacture without having to measure accurately the many parameters necessary for use in (4.14).

To illustrate a typical jitter tolerance template for a regenerator, we note from

Figure 4.7 that a sinusoidal alignment jitter of $81°$ peak-to-peak results in a BER penalty of 0.5 dB when $0°$ static phase offset and a raised cosine pulse shape was assumed. Inserting $e_{aN}^{P-P} = 81°$ into (4.19) and assuming a second-order Butterworth $H_N(f_{pm})$ of magnitude

$$|H_N(f_{pm})| = \frac{1}{\sqrt{\left(\frac{f_{pm}}{B}\right)^4 + 1}} \qquad (4.20)$$

and linear argument

$$\arg H_N(f_{pm}) = \phi_P f_{pm} \qquad (4.21)$$

where ϕ_P is a linear phase slope, we graph the jitter tolerance template in Figure 4.8. Figure 4.8 shows the expected jitter tolerance template of a regenerator with an ideal second-order Butterworth jitter transfer function, a raised cosine pulse shape, and a zero degree static phase offset. The template assumes a 0.5 dB BER penalty, $\tau = 0$ and $g_N(t)$ given by (4.15).

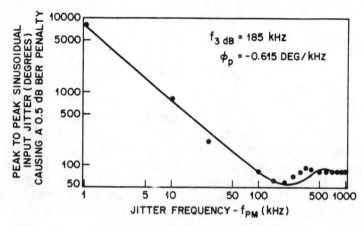

Fig. 4.8 Comparison of a measured and a calculated jitter tolerance curve for a fiber optic regenerator

We will now discuss a jitter tolerance measurement procedure and results of measurements made on a fiber optic regenerator during manufacture [4.11]. Figure 4.9 shows the jitter tolerance measurement configuration. A signal generator generates a sine wave of amplitude A and frequency f_{pm}. This sine wave is used to frequency modulate a clock at the baud of the system, here 295.6 MHz. This frequency modulated clock then reads out a shift register data sequence with period $2^{23} - 1$ bits from a data generator. The peak-to-peak amplitude of the sinusoidal jitter at frequency f_{pm} on the data signal is given by [4.12]

$$2K_i = \frac{m_{fm}A}{f_{pm}} \frac{360°}{2\pi} \qquad (4.22)$$

where m_{fm} is the frequency modulation index of the modulated clock in volts/kHz.

The data signal is sent optically by the laser transmitter through an optical attenuator or a long fiber span to the regenerator's receiver. With the sinusoidal jitter initially set to zero, the optical attenuator is set such that the regenerator is performing at a BER of 1×10^{-9}. Next the optical attenuation is decreased 0.5 dB so that the SNR at the decision circuit is 0.5 dB better and the regenerator is performing with a P_E of better than 1×10^{-9}. The jitter tolerance is measured by increasing A until the P_E returns to 1×10^{-9} and recording $2K_i$. This procedure is then repeated at various f_{pm}. The graph of $2K_i$ vs. f_{pm} is the amount of peak to peak sinusoidal input jitter resulting in a BER penalty of 0.5 dB.

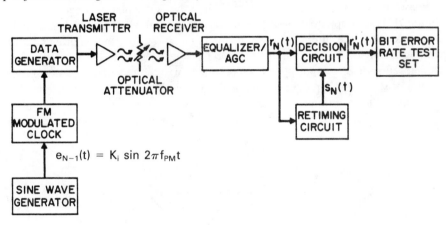

Fig. 4.9 Jitter tolerance measurement configuration

Figure 4.8 shows results of a jitter tolerance measurement made on a fiber optic regenerator along with the calculated curve. The measured values follow along a curve that is similar in shape to the calculated jitter tolerance curve, but differs slightly at certain values. This difference is caused by a nonzero static phase offset inside the regenerator under test. Next we will show how jitter tolerance can be used to help eliminate this static phase offset, and to improve the regenerator for long-haul system use.

Thus far, much attention has been given to the effect that a static phase offset has on P_E. Previously we showed that the effect of alignment jitter was more harmful when significant static phase offsets were present. Since the timing circuit is a narrowband device, with a usual delay of several hundred pulse periods [4.13], manufacturing variations in timing circuit delays result in an arbitrary timing signal phase relative to the data signal. To assure that the rising zero crossing of the timing signal occurs at the most optimum time in the pulse period, a variable length coaxial cable or a tunable electronic circuit is commonly used to set up the optimum phase relationship. We will show that selecting the phase relationship using a jitter tolerance criteria will improve the regenerator's overall jitter performance.

Previously we showed that the BER penalty increased when alignment jitter and a static phase offset were both present. Since jitter tolerance is a measure of BER penalty in the presence of alignment jitter and static phase offset, jitter tolerance will be maximized when the phase relationship between the timing and data signals is optimum. Conversely, finding the phase relationship that maximizes the jitter tolerance improves the regenerator's jitter performance. We have used this technique to find the optimum delay line length for a fiber optic regenerator.

Figure 4.10 shows results of jitter tolerance measurements made at $f_{pm} = 100$ kHz for various delay line lengths along with the measured BER penalty in the presence of zero input jitter for each delay line length chosen. Figure 4.10 shows that a decrease in jitter tolerance is measured when the static phase offset is ±5° from optimum. It is also apparent from Figure 4.10 that a delay line length equal to 10.3 inches maximizes the jitter tolerance of this regenerator and thus improves its jitter performance. Also, we see that choosing a delay line length based on BER penalty in the presence of zero input jitter does not maximize the regenerator's performance in a system where an appreciable amount of accumulated jitter is expected. Therefore, from a jitter standpoint, using a jitter tolerance criteria to select the optimum sampling time inside the regenerator improves the regenerator's tolerance to input jitter.

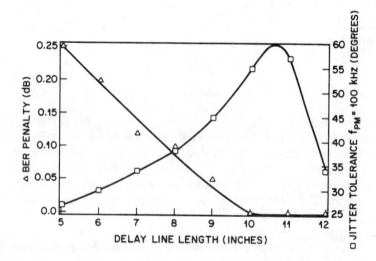

Fig. 4.10 Comparison of the jitter tolerance criterion *versus* a BER penalty criterion for optimizing the static phase offset of a fiber optic regenerator

4.4 ALIGNMENT JITTER ACCUMULATION

In Chapter 3, we modeled the accumulation of random jitter and systematic jitter. We found the rms and the power spectrum of both the random and systematic accumulated jitter of a long chain of regenerators. This accumulated jitter is referenced to the clock of the nonjittered pulse pattern that was sent from the multiplexer to the first regenerator of the system, and is only important at the demultiplexing end of the transmission system where the data signal must be sent out to other parts of the synchronized digital network. There, the jittered data signal must again be synchronized to a standard "unjittered" clock, by entering an elastic store that is then clocked out by the standard reference clock (see Chapter 5). The magnitude of the accumulated jitter determines the size of the elastic store needed. In this chapter we have shown that it is the magnitude of the alignment jitter that will effect system performance. Alignment jitter is the local deviation of the timing signal's zero crossing from the center of the pulse period. To relate the jitter accumulation models of Chapter 3 to the alignment jitter at the Nth regenerator, we separate the alignment jitter into random and systematic components.

The random alignment jitter at the Nth regenerator is

$$e_{aN}^{R}[nT] = e_{N}^{R}[nT] - e_{N-1}^{R}[nT] \tag{4.23}$$

Substituting the expression for $e_{N}^{R}[nT]$ given by (3.1) into (4.23) we have

$$e_{aN}^{R}[nT] = (e_{N-1}^{R}[nT] + e_{iN}^{R}[nT]) * h_{N}[nT] - e_{N-1}^{R}[nT] \tag{4.24}$$

Rewriting (4.24) as

$$e_{aN}^{R}[nT] = e_{N-1}^{R}[nT] * [h_{N}[nT] - \delta[nT]] + e_{iN}^{R}[nT] * h_{N}[nT] \tag{4.25}$$

and recalling that $e_{iN}^{R}[nT]$ is uncorrelated with $e_{N-1}^{R}[nT]$ by definition, results in the power spectrum of $e_{aN}^{R}[nT]$ of [4.14]:

$$\Phi_{aN}^{R}(f) = \Phi_{N-1}^{R}(f)|H_{N}(f) - 1|^{2} + \Phi_{iN}^{R} |H_{N}(f)|^{2} \tag{4.26}$$

The rms of $e_{aN}[nT]$ is

$$\sigma_{aR}[N] = \sqrt{\int_{-1/T}^{1/T} \Phi_{aN}^{R}(f)df} \tag{4.27}$$

To illustrate the properties of random alignment jitter accumulation, we simplify (4.26) by assuming identical regenerators. For identical regenerators $\Phi_{N-1}^R(f)$ is given by (3.12) and simplifies (4.26) to

$$\Phi_{aN}^R(f) = \Phi_{i1}^R |H_1(f)|^2 \frac{1 - |H_1(f)|^{2N-2}}{1 - |H_1(f)|^2} |H_1(f) - 1|^2 + \Phi_{i1}^R |H_1(f)|^2 \quad (4.28)$$

Assuming a second order low-pass filter for $H_1(f)$ given by (3.16), we plot the normalized random rms alignment jitter accumulation, $\sigma_{aR}[N]/\sigma_{aR}[1]$ for various amounts of jitter peaking. Figure 4.11 shows that with increasing jitter peaking, $\sigma_{aR}[N]$ accumulates exponentially. Therefore, when many regenerators are cascaded, the P_E in the latter regenerators in the chain is degraded because of increasing random alignment jitter. Depending on τ and $g_N(t)$, the penalty can be significant if $\sigma_{aR}[N]$ is allowed to increase rapidly.

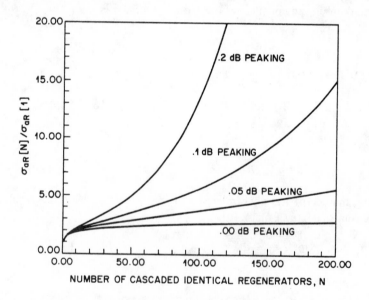

Fig. 4.11 Normalized rms random alignment jitter accumulation

The systematic alignment jitter at the Nth regenerator is

$$e_{aN}^S[nT] = e_N^S[nT] - e_{N-1}^S[nT] \quad (4.29)$$

Substituting (3.5) for $e_N^S[nT]$ and $e_{N-1}^S[nT]$, we have

$$e_{aN}^S[nT] = e_{i1}^S[nT] * h_1[nT] * h_2[nT] * \ldots * h_N[nT]$$
$$- e_{i1}^S[nT] * h_1[nT] * \ldots * h_{N-1}[nT]$$
$$+ e_{i2}^S[nT] * h_2[nT] * \ldots * h_N[nT]$$
$$- e_{i2}^S[nT] * h_2[nT] * \ldots * h_{N-1}[nT]$$
$$\cdot$$
$$\cdot$$
$$\cdot$$
$$+ e_N^S[nT] * h_N[nT] \tag{4.30}$$

Since the systematic jitter generated by a particular regenerator is completely correlated with the systematic jitter generated by the other regenerators, the power spectrum of $e_{aN}^S[nT]$ is

$$\Phi_{aN}^S(f) = \left| \sqrt{\Phi_{i1}^S} \left(\prod_{l=1}^N H_1(f) - \prod_{l=1}^{N-1} H_1(f) \right) \right.$$
$$\left. + \sqrt{\Phi_{i2}^S} \left(\prod_{l=2}^N H_1(f) - \prod_{l=2}^{N-1} H_1(f) \right) + \ldots \sqrt{\Phi_{iN}^S} \, H_N(f) \right|^2 \tag{4.31}$$

The rms of $e_{aN}^S[nT]$ is

$$\sigma_{aS}^S[N] = \sqrt{\int_{-1/T}^{-1/T} \Phi_{aN}^S(f) df} \tag{4.32}$$

To illustrate the properties of systematic alignment jitter, we shall assume identical regenerators. Substituting (3.15) and (4.32) simplifies (4.32) to

$$\Phi_{aN}^S(f) = \Phi_{i1}^S |H_1^N(f)|^2 \tag{4.33}$$

Again assuming a second order $H_1(f)$, we graph $\sigma_{aS}[N]/\sigma_{aS}$ [1] for various amounts of jitter peaking [4.15] in Figure 4.12. From (4.33) and Figure 4.12, if $H_1(f)$ shows no jitter peaking, then the systematic alignment jitter decreases along the chain. If there is jitter peaking $\sigma_{aS}[N]$ increases exponentially, however not as rapidly as $\sigma_{aR}[N]$.

From these examples, we see that alignment jitter accumulation is dominated by the random type. With no jitter peaking, $\sigma_{aS}[N]$ decreases as the number of cascaded identical regenerators, while $\sigma_{aR}[N]$ accumulates exponentially for long systems. With jitter peaking, $\sigma_{aR}[N]$ and $\sigma_{aS}[N]$ accumulate exponentially with $\sigma_{aR}[N]$

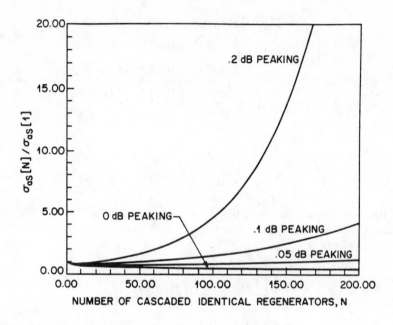

Fig. 4.12 Normalized rms systematic alignment jitter accumulation

quickly dominating. We again find that jitter peaking in $|H_N(f)|$ must be controlled to prevent alignment jitter from degrading the error performance of the digital transmission system. Therefore, proper design of the timing extraction circuit is necessary to assure that transmission penalty caused by jitter are minimized (see Chapter 2).

4.5 REMARKS

In this chapter, we have studied the effect accumulated alignment jitter has on the bit decisions inside the Nth cascaded regenerator. We then defined the jitter tolerance of a regenerator as an effective measure to assure that excessive transmission penalty will not occur when the regenerator is placed at the end of a long chain of regenerators. We also showed that the optimum sampling time inside the regenerator can be found effectively using a jitter tolerance technique. Finally we modeled alignment jitter accumulation using the nonidentical and identical regenerator models of Chapter 3, so that the alignment jitter defined in this chapter could be related to the overall jitter accumulation.

Thus far in this book we have studied the details of jitter introduced by line regenerators. The concept explored in Chapters 2 through 4 are general enough that they can usually be applied whenever timing extraction or data sampling is used. We will now turn our attention on the jitter introduced by multiplex equipment.

REFERENCES

[4.1] B. R. Saltzberg, "Timing Recovery for Synchronous Binary Data Transmission," *Bell System Technical Journal*, March 1967.

[4.2] K. Trondle and G. Soder, *Optimization of Digital Transmission Systems*, Artech House, 1987.

[4.3] R. W. Lucky, J. Salz, and E. J. Weldon, *Principles of Data Communications*, McGraw-Hill, 1968.

[4.4] V. K. Prabhu, "Some Considerations of Error Bounds in Digital Systems," *Bell System Technical Journal*, 1971, Vol. 50, pp. 3127–3151.

[4.5] T. Aratani, "Jitter Effect on the Error Rate of a PCM Repeater," *Electronics and Communications Journal of Japan*, November 1966.

[4.6] H. Nyquist, "Certain Factors Affecting Telegraph Speed," *Bell System Technical Journal*, Vol. 3, April 1924.

[4.7] A. Papoulis, *Probability, Random Variables, and Stochastic Processes*, McGraw-Hill, 1965.

[4.8] C. C. Cook, "Jitter Tolerances in Digital Equipment," *IEE Colloquium on Jitter in Digital Communications Systems*, London, 1977.

[4.9] R. D. Hall and M. J. Snaith, "Jitter Specifications in Digital Networks," *Proceedings of EEJ*, Vol. 72, July 1979.

[4.10] P. R. Trischitta and P. Sannuti, "The Jitter Tolerance of Fiber Optic Regenerators," *IEEE Transactions on Communications*, Vol. 35, No. 12, December 1987.

[4.11] D. G. Ross, R. M. Paski, D. G. Ehrenberg and G. M. Homsey, "A Highly Integrated Regenerator for 295.6 Mb/s Undersea Optical Transmission," *IEEE/OSA Journal of Lightwave Technology*, Vol. LT-2, No. 6, December 1984.

[4.12] Transmission Systems for Communications, Fifth Edition, Bell Telephone Laboratories, 1982.

[4.13] R. L. Rosenberg, C. Chamzas and D. A. Fishman, "Timing Recovery with SAW Transversal Filters in the Regenerators of Undersea Long-Haul Fiber Transmission Systems," *IEEE/OSA Journal of Lightwave Technology*, Vol. LT-2, No. 6, December 1984.

[4.14] D. A. Fishman, R. L. Rosenberg, C. Chamzas, "Analysis of Jitter Peaking Effects in Digital Long-Haul Transmission Systems Using SAW Filter Retiming," *IEEE Transactions on Communications*, Vol. COM-33, No. 7, July 1985.

[4.15] E. Roza, "Analysis of Phase Locked Timing Extraction Circuits for Pulse Code Transmission," *IEEE Transactions on Communications*, Vol. COM-22, No. 9, September 1979.

Chapter 5
Jitter Introduced by Digital Multiplexes

Multiplexing is the process whereby several lower-rate signals are interleaved to form a single, composite, higher-rate signal for transmission over a single medium. Digital multiplexing accomplishes this by employing the principle of time-sharing. Each of the lower-rate signals time-share the higher-rate signal, so that all inputs are combined into a single output. Conversely, demultiplexing is the process whereby the lower-rate signals are separated from the composite higher-rate signal. A fundamental problem of digital multiplexing is accommodating signals that are not identical in bit rate. It is necessary to equalize the rates of the signals to be interleaved for transmission over a single medium. The process of bit rate equalization is termed *synchronization*. Major issues in digital multiplex system design relate to mechanisms for synchronizing the input signals so that they can be properly interleaved, and to methods for dealing with the jitter that is generated on demultiplexing.

Conceptually, time-division multiplexing of several digital signals, into a single higher-rate signal, can be done within a multiplexing terminal by a commutator-like selector switch that takes a bit from each lower-rate input in turn, and applies it to the higher-rate output (see Figure 5.1). At the demultiplexer, the inverse operation is performed. The higher-rate signal is thereby separated into its component parts, thus recovering the lower-rate inputs. A simple example of time-division multiplexing is given in Figure 5.2 [5.1]. In Figure 5.2a the outputs of two signals are combined in the multiplexer into a single composite higher-rate signal. If the two signals have precisely the same bit rate, their relative phase is constant (shown in Figure 5.2b), and the composite output signal is as shown in Figure 5.2c. However, if the input signal rates are not identical, the relative phase between the two signals is time varying. Furthermore, even with input signals that initially have identical rates, slow variations in the propagation delay of the transmission medium connecting the input signals and the multiplexer, for example, result in phase variations between the input signals (see Chapter 8). Then the bits at point C (in Figure 5.2a) would be jittered, and if severe, the two sets of input signal bits in the composite signal would be indistinguishable, as shown in Figure 5.2d.

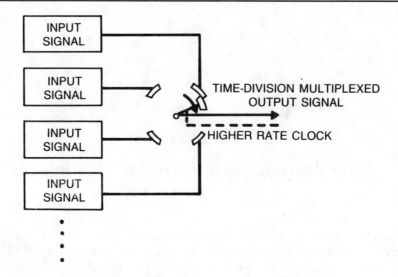

Fig. 5.1 Time-division multiplexing concept

A digital multiplex (multiplexer-demultiplexer pair) may be categorized in terms of the synchronization scheme it uses to equalize the bit rates of the incoming signals. The primary synchronization schemes in use today are slip buffering, bit justification, and pointer processing. In this chapter we will discuss these synchronization schemes and how each generates jitter.

5.1 MULTIPLEXING USING SLIP BUFFERING

We will begin with slip buffering synchronization because the generated jitter is simplest to analyze. The primary multiplex components affecting jitter are the lower- and higher-rate timing extraction circuits, the multiplexer output timing signal, and the synchronizers and desynchronizers. In slip buffering synchronization each incoming lower-rate signal enters a synchronizer, where an elastic store is used to equalize its bit rate to a common rate before bit interleaving. Bits are written into the elastic store using a timing signal extracted from the incoming lower-rate signal (Chapter 2), and bits are read out of the store using a common local timing signal. Phase variations between the read and write timing signals are absorbed by the elastic store. Acceptable operation of a multiplexer using slip buffering synchronization places constraints on the absolute phase difference between the read and write timing signals. Specifically, these timing signals must either be synchronous or plesiochronous with respect to one another.

Synchronous signals are defined as having the same long-term frequency, which implies a bounded *phase difference*. This requires some sort of ongoing communication of timing information among synchronous sources. Plesiochronous signals are

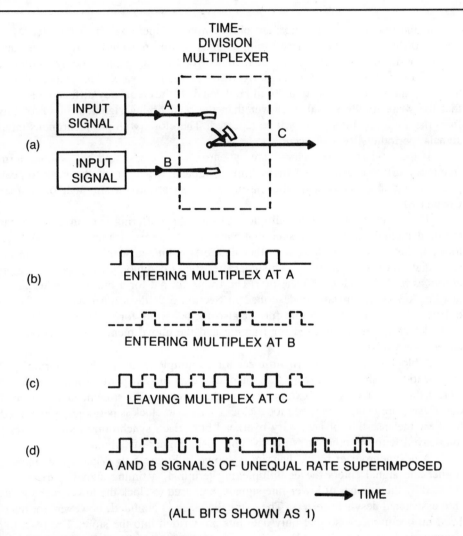

TIME-
DIVISION
MULTIPLEXER

(a)

INPUT SIGNAL

INPUT SIGNAL

A

B

C

(b) ENTERING MULTIPLEX AT A

ENTERING MULTIPLEX AT B

(c) LEAVING MULTIPLEX AT C

(d) A AND B SIGNALS OF UNEQUAL RATE SUPERIMPOSED

→ TIME

(ALL BITS SHOWN AS 1)

Fig. 5.2 Example of time-division multiplexing

defined as having a tightly bounded *frequency difference*. This is physically achievable without timing information communication. For example, consider two clocks that have different intrinsic frequency departures from a nominal value. In one case, the second clock is phase-locked to the first; timing outputs from these two clocks are now synchronous with respect to each other. In the second case, the two clocks are free-running. If the frequency departure for each of these clocks is constrained to be within one part in 10^{11} (referred to as a "Stratum 1" clock), timing outputs from these two clocks are considered plesiochronous. Note that if the frequency departures for each of these clocks are greater than one part in 10^{11} they are considered asynchronous.

If the read and write clocks are plesiochronous, there will be a long-term frequency difference between these timing signals. If the frequency difference is such that the write timing signal is faster than the read timing signal, the synchronizer elastic store will fill. When the store overflows, a "slip" results, and a block of data bits from the incoming bit stream will be deleted. If the frequency difference is such that the write timing signal is slower than the read timing signal, a block of bits from the incoming bit stream will be repeated. Therefore, with plesiochronous input signals, periodic slips will occur.

If the read and write clocks are synchronous, there should be[1] no long-term frequency difference between these timing signals. However, if the peak-to-peak phase variation exceeds what the elastic store can absorb, a slip will occur (see Chapter 8).

The frame format for a multiplexer using slip buffering synchronization has the fundamental characteristic such that once the multiplexer framing bits have been located, the location of all of the other data bits is known with no need for an intermediate level of processing. This characteristic is commonly described as a *synchronous multiplexing* characteristic. (The frame format for a multiplexer using bit justification synchronization, described in Section 5.2, does not have this characteristic.) The multiplexer frame format also includes bits for signaling, telemetry, error detection, etc. Bits used for framing and the above mentioned functions are called overhead bits.

A block diagram of a synchronizer for a multiplexer using slip buffering synchronization is shown in Figure 5.3. Incoming data bits are written into the elastic store using a timing signal extracted from the incoming bit stream, and then read out of the store using a common local clock. This read clock is periodically inhibited to allow the insertion of necessary overhead bits. Each synchronized signal is then interleaved with the others to form the composite higher-rate signal.

A desynchronizer block diagram is shown in Figure 5.4. When the composite higher-rate signal enters the demultiplexing terminal, a timing signal is extracted, divided by the number of lower-rate signals, and used to clock the lower-rate signals into associated desynchronizers. The extracted clock is inhibited, however, for overhead bit occurrences, so that only data bits are written into the store. The resulting gapped clock may be filtered by a phase-smoothing circuit. Because of the low-pass nature of phase-smoothing circuit implementations, some jitter remains on the outgoing lower-rate signal. An alternative to the phase-smoothing circuit is to use a local synchronous timing signal to read the data from the elastic store. Then, no jitter from overhead bit deletion would be present on the outgoing lower-rate signal. However, periodic slips could result.

Overhead bit deletion causes a periodic sawtooth-like jitter to be generated. Let us define M as the number of lower-rate input signals, and m as the number of times

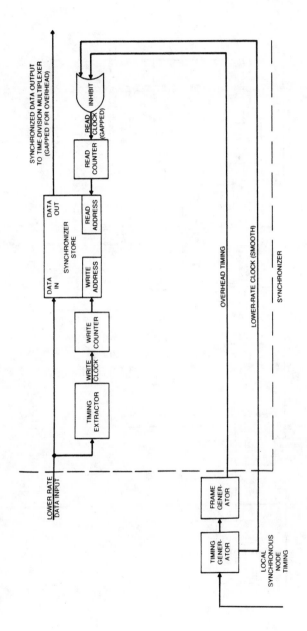

Fig. 5.3 Slip buffering multiplex: synchronizer block diagram

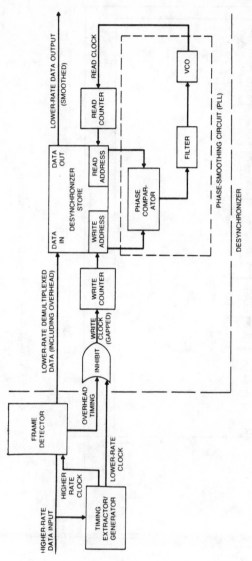

Fig. 5.4 Slip buffering multiplex: desynchronizer block diagram

a bit from one of the lower-rate input signal appears between overhead bits in the multiplexer frame format. Then, if M unjittered lower-rate input signals are bit interleaved, with one overhead bit added every $m \times M$ bits, the multiplexer output frequency will be given by

$$f_{out} = \frac{M}{T} \frac{(mM + 1)}{mM} = \frac{(mM + 1)}{mT} \tag{5.1}$$

where T is the input bit period.

The addition of the overhead bit will effectively add a sawtooth jitter waveform, $e_{oh}(t)$, of approximate amplitude $1/M$ and period mT to each lower-rate signal (see Figure 5.5). We can represent $e_{oh}(t)$ by its Fourier series where we ignore the dc term

$$e_{oh}(t) = \sum_{n=1}^{\infty} a_n \cos\left(\frac{2\pi nt}{mT}\right) + \sum_{n=1}^{\infty} b_n \sin\left(\frac{2\pi nt}{mT}\right) \tag{5.2}$$

where

$$a_n = \frac{2}{mT} \int_{-mT/2}^{mT/2} \frac{1}{M} \left(\frac{t}{mT} + \frac{1}{2}\right) \cos\left(\frac{2\pi nt}{mT}\right) dt = 0 \tag{5.3a}$$

and

$$b_n = \frac{2}{mT} \int_{-mT/2}^{mT/2} \frac{1}{M} \left(\frac{t}{mT} + \frac{1}{2}\right) \sin\left(\frac{2\pi nt}{mT}\right) dt = \frac{(-1)^{n+1}}{\pi nM} \tag{5.3b}$$

Thus,

$$e_{oh}(t) = \frac{1}{\pi M} \sum_{n=1}^{\infty} \frac{(-1)^{n+1}}{n} \sin\left(\frac{2\pi nt}{mT}\right) \tag{5.4}$$

Using a phase-smoothing circuit to smooth the output data from the demultiplexer, the jitter on the demultiplexer output signal is

$$e_D(t) = g(t) * e_{oh}(t) \tag{5.5}$$

where $g(t)$ is the impulse response of the phase-smoothing circuit.

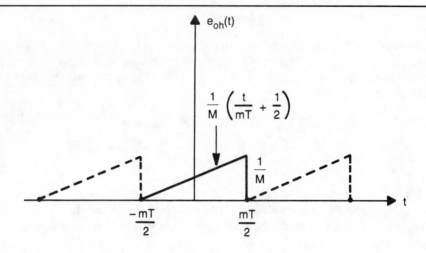

$e_{oh}(t)$ AS A FUNCTION OF TIME

Fig. 5.5 $e_{oh}(t)$ as a function of time

In typical applications, the phase-smoothing circuit bandwidth, B, will be much less than the overhead repetition rate:

$$B \ll \frac{1}{mT} \qquad (5.6)$$

Thus, if the phase-smoothing circuit is a second-order phase-locked loop with a large damping factor (e.g., $\xi = 5$),

$$G(f) \approx \frac{-jB}{f} \qquad (5.7)$$

where $G(f)$ is the transfer function of the phase-smoothing circuit and $f \gg B$ is the range of applicability. Then, the output jitter signal is given by

$$e_D(t) \approx F^{-1}\left\{ \frac{-jB}{f} \cdot \frac{1}{\pi M} \sum_{n=1}^{\infty} \frac{(-1)^{n+1}}{n} \cdot \frac{1}{2j} \left\{ \delta\left(f - \frac{n}{mT}\right) - \delta\left(f + \frac{n}{mT}\right) \right\} \right\}$$

$$\approx \frac{B}{2\pi M} \int_{-\infty}^{\infty} \sum_{n=1}^{\infty} \frac{(-1)^{n+1}}{n} \frac{e^{j2\pi ft}}{f} \left\{ \delta\left(f + \frac{n}{mT}\right) - \delta\left(f - \frac{n}{mT}\right) \right\} df \qquad (5.8)$$

$$\approx \frac{BmT}{\pi M} \sum_{n=1}^{\infty} \frac{(-1)^n}{n^2} \cos\left(\frac{2\pi nt}{mT}\right) \qquad (5.9)$$

with mean square amplitude of

$$
\begin{aligned}
\sigma_D^2 &= \left(\frac{BmT}{\pi M}\right)^2 E\left\{ \sum_{n=1}^{\infty} \frac{(-1)^n}{n^2} \cos \frac{2\pi nt}{mT} \sum_{l=1}^{\infty} \frac{(-1)^l}{l^2} \cos \frac{2\pi lt}{mT} \right\} \\
&= \left(\frac{BmT}{\pi M}\right)^2 \sum_{n=1}^{\infty} \frac{1}{n^4} E\left\{ \cos^2 \frac{2\pi nt}{mT} \right\} \\
&= \left(\frac{BmT}{\pi M}\right)^2 \frac{\pi^4}{90} \cdot \frac{1}{2}
\end{aligned}
\tag{5.10}
$$

and rms amplitude, σ_D, of

$$
\sigma_D \approx 0.74 \left\{ \frac{BmT}{\pi M} \right\}
\tag{5.11}
$$

5.2 MULTIPLEXING USING BIT JUSTIFICATION

A more commonly used mechanism for synchronizing the lower-rate input signals is bit justification, also called bit stuffing. The term *justification* originated in the printing industry where it describes the process of adjusting the spaces between printed words so that all the lines of print have the same length. Another practical example of justification is embodied by the concept of leap year. A nominal length calendar year is 365 days long, but to make the calendar year nearly the same as the solar year, an extra day is added to the year at the end of February once every four years (i.e., every 4 × 365 days, a day is justified).

Three bit justification synchronization schemes have been used: (1) positive justification, (2) negative justification, and (3) positive-negative and positive-zero-negative justification. Although some discussion of the latter two schemes is included within this chapter for completeness, we will focus on the positive justification scheme, as it is most commonly used.

Figure 5.6 shows the primary multiplex functional components affecting jitter: the lower- and higher-rate timing extraction circuits, the multiplexer output timing signal, and the synchronizers and desynchronizers. With positive bit justification synchronization, each incoming signal enters a synchronizer where extra bits are inserted as often as required to equalize its bit rate to a specific common rate before bit interleaving. The justified bits, which carry no information, are removed within desynchronizers when the composite higher-rate signal is demultiplexed. As may be inferred from the above discussion, the rate of the outgoing composite signal is slightly higher than the sum of the maximum rates of the incoming lower-rate signals (e.g., $f_{out} > Mf_{in}$). Positive bit justification operation within a synchronizer may be described using the commutator analogy. Referring to Figure 5.7 [5.2], one brush writes

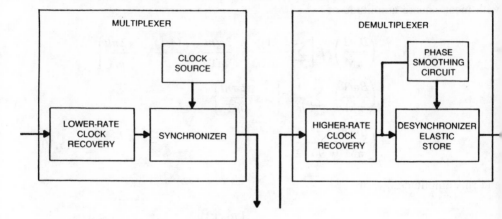

Fig. 5.6 Multiplex functional components affecting jitter

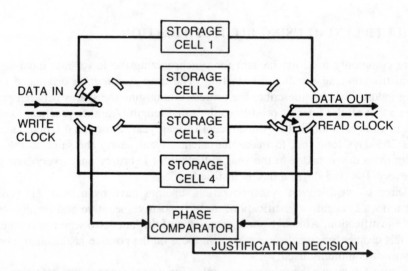

Fig. 5.7 Justification concept

the data into elastic storage cells, and a second brush reads the data out. The angular velocities of the brushes correspond to the frequencies of the write and read clocks. When the input signal rate is slower than the synchronized output data rate, the commutator read brush begins to catch up to the commutator write brush. If the read brush were allowed to overtake the write brush, data would be read twice (a slip would occur). To avoid this, a phase comparator monitors the phase difference between the brushes. If a slip is imminent, the read brush is stopped for one read clock pulse to allow the write brush to catch up.

Implementation of the positive bit justification process within digital multiplexes involves assigning bit positions within the multiplexer frame format on a per-input-signal basis that are used for: (1) signaling a positive justification decision, and (2) positive justification opportunities. It is necessary for justification opportunities to occur only in identifiable bit positions. Equally important is the capability to signal the receiving demultiplexer that a justification decision has been made, and that the next assigned bit position will be justified. (The multiplex frame format also includes the overhead bits described in the previous section.)

An example of a positive bit justification multiplexer frame format is given below to show the basic principles and terminology which will be used in the remainder of this section. The example format is for a DS1 (f_{nom} = 1.544 MHz) to DS2 (f_{out} = 6.312 MHz) multiplexer, which takes M = 4 DS1 input signals, synchronizes them to a common rate using positive bit justification, and then bit interleaves them to form a DS2 output signal. Referring to Figure 5.8 [5.2], the frame format is comprised of a periodic pattern, repeating twenty-four times, of a single overhead bit (an M, C, or F bit) followed by a sequence of twelve sets of four bit-interleaved, rate-equalized, DS1 input signals. Thus, the frame length is [24 × (1 + 12 × 4)] = 1176 DS2 bits. Denoting the four DS1 input signals by I_I, I_{II}, I_{III}, and I_{IV}, respectively, every bit from the I_I input is normally separated by four DS2 signal unit intervals, but every twelfth bit from the I_I input is separated from its successor by five DS2 signal unit intervals. The same holds for the other three DS1 inputs. Note that once per frame, each input has a justification opportunity. Therefore, the maximum rate that justification can occur is once per 1176 DS2 unit intervals. Note that once the framing bits have been located, the location of all the other data bits is *not* known without further processing (e.g., to determine whether a particular bit position contains data or has been justified).

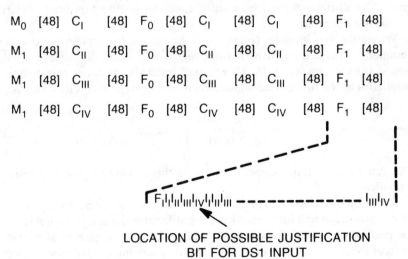

LOCATION OF POSSIBLE JUSTIFICATION
BIT FOR DS1 INPUT

Fig. 5.8 DS1 to DS2 multiplexer format

Figures 5.9 and 5.10 illustrate functional block diagrams of a multiplex synchronizer and desynchronizer, respectively. In the synchronizer data bits are written into an elastic store using an extracted timing signal of rate $f_{nom}{}^2$, and are read out of the store at the synchronized output clock rate, f_r. The synchronized output clock rate is the local multiplexer clock rate, f_{out}, divided by the number of lower-rate input signals, M. The synchronizer phase comparator outputs the phase difference between the read and write clocks, which is monitored by control circuitry. When the synchronizer phase comparator output shows that the store is becoming depleted (i.e., the phase difference between the read and write clocks has reached the justification decision threshold) a read clocking bit is inhibited at the next justification opportunity, causing the phase difference between the read and write clocks to decrease by f_{nom}/f_r of an elastic store cell. Note that the read clock is also inhibited when overhead bits are inserted into the outgoing bit stream.

In a demultiplexer, the write clock is extracted from the incoming signal (using techniques discussed in Chapter 2), divided by M, and used to clock the lower-rate signals into desynchronizers. Within a desynchronizer, this clock is inhibited (gapped) for overhead and justification bit occurrences so that only data bits are written into the store. The desynchronizer phase-smoothing circuit generates a clock having the same long-term average frequency as this gapped clock, which is used to read the data bits out of the store. However, as phase-smoothing circuits have a low-pass nature, low frequency jitter is present on the outgoing lower-rate signals.

It should be emphasized that the generated jitter occurs not solely from the justification process, but also because of the waiting time between justification decision threshold crossing and the justification opportunity. This inability to justify on demand causes the phase difference between the read and write clocks to continue to grow. The distinction between waiting time and justification jitter is clarified in the following discussion [5.4].

Within the synchronizer (Figure 5.9) the bits on the synchronized data output are normally separated by M higher-rate signal unit intervals, but every mth bit is separated from its successor by M + 1 unit intervals, to allow an overhead bit to be inserted into the higher-rate signal. Then, the average read clock rate is

$$\bar{f}_r = \frac{f_{out}}{M} \times \frac{mM}{(mM + 1)} = f_{out} \times \frac{m}{(mM + 1)} \tag{5.12}$$

For the remainder of this section, the jitter on this clocking signal will be neglected for simplicity.

Referring to Figure 5.11, $e_{SPC}(t)$ is the output of the synchronizer phase comparator expressed in unit intervals, Λ is the justification decision threshold, and $e_{WT}(t)$ is the portion of $e_{SPC}(t)$ exceeding Λ (e.g., waiting time jitter). Note that in the absence of input jitter, the jitter on the gapped synchronizer read clock output, $e_s(t)$, is equivalent to the synchronizer phase comparator output. The time between justification opportunities is t_j, and the time between the ith preceding justification op-

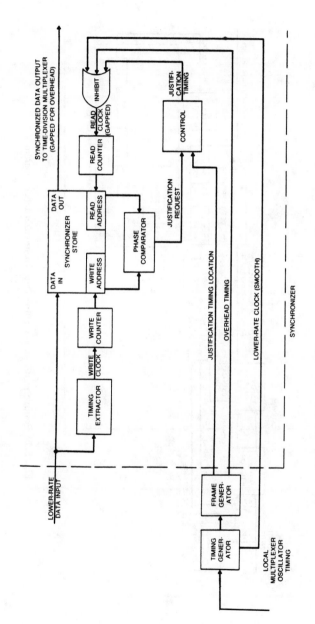

Fig. 5.9 Positive justification multiplex: synchronizer block diagram

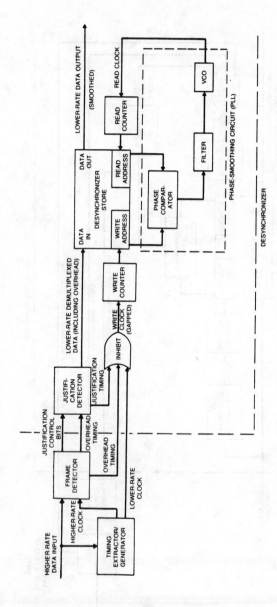

Fig. 5.10 Positive justification multiplex: desynchronizer block diagram

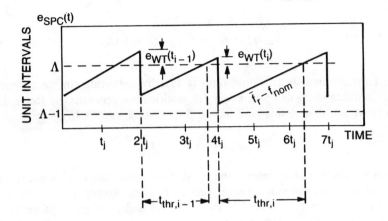

Fig. 5.11 $e_{WT}(t)$ as a function of time

portunity and the ith justification decision threshold crossing is $t_{thr,i}$.

Figure 5.11 shows the output of the synchronizer phase comparator, which is a series of sawtooths, and the deviation of $e_{WT}(t)$ with respect to time. From Figure 5.11 we note that

$$e_{WT}(t_i) + (\bar{f}_r - f_{nom})t_{thr,i} = 1 \text{ UI} \qquad (5.13)$$

The difference frequency between the read and write clocks is an important relationship, and is defined as the nominal justification rate. The nominal justification ratio, ρ, is defined as the ratio of the nominal justification rate, $f_{j,nom}$, and the maximum justification rate $f_{j,max}$.

$$\rho \equiv \frac{f_{j,nom}}{f_{j,max}} = \frac{\bar{f}_r - f_{nom}}{f_{j,max}} \qquad (5.14)$$

From one justification opportunity to another, the waiting time jitter is

$$e_{WT}(t_{i-1}) + (\bar{f}_r - f_{nom})nt_j = e_{WT}(t_i) + 1 \text{ UI} \qquad (5.15)$$

where n is an integer.
Consider the case in which the justification ratio is the inverse of an integer. Then,

$$\rho \equiv \frac{1}{j} = \frac{\bar{f}_r - f_{nom}}{f_{j,max}} = (\bar{f}_r - f_{nom})t_j$$

and (5.15) may be expressed as

$$e_{WT}(t_{i-1}) + \frac{1}{j} n = e_{WT}(t_i) + 1 \text{ UI} \tag{5.16}$$

However, if the nominal justification ratio is $1/j$, by definition (in the absence of input jitter) justification will occur every jth justification opportunity. Referring to Figure 5.11, we see that in this case $n = j$. Substituting into (5.16),

$$e_{WT}(t_{i-1}) = e_{WT}(t_i) \tag{5.17}$$

Thus, if ρ is the inverse of an integer the waiting time jitter does not impress a low-frequency envelope on the synchronizer phase comparator output, as illustrated in Figure 5.12a. Figures 5.12b and 5.12c show that for ρ slightly greater than $1/2$ and ρ slightly less than $1/2$ a strong low-frequency envelope is impressed on the synchronizer phase comparator output. In general, whenever ρ is very close to a simple rational number (a rational number with a small denominator), appreciable low-frequency power will be present [5.6].

From the above discussion, it is apparent that the jitter generated by justification is of much higher frequency components than that generated by waiting time. Although the low-pass filtering action of the desynchronizer phase-smoothing circuit usually eliminates the justification jitter, waiting time jitter frequency components are typically low enough to fall within the phase-smoothing circuit passband.

We will now give an example of the maximum waiting time jitter that may result for a given frame format [5.4]. The maximum amplitude of waiting time jitter is equal to the maximum difference frequency, $\bar{f}_r - f_{nom}$, multiplied by t_j. As discussed earlier, in the DS1 to DS2 multiplexer format, there are 1176 DS2 unit intervals between justification opportunities. The maximum allowable DS1 and DS2 signal tolerances are ± 130 ppm and ± 33 ppm, respectively [5.3]. Therefore, the maximum difference frequency is obtained by using the maximum average data bit read clock frequency and the minimum write clock frequency. As the maximum DS2 signal frequency is 6.312208×10^6 Hz, then using (5.12), the maximum average read clock frequency is

$$\bar{f}_{r+} = 6.312208 \times 10^6 \times \frac{12}{49} = 1.5458469 \times 10^6 \text{ Hz}$$

The time between justification opportunities at this DS2 data rate is

$$t_j = \frac{1}{(6.312208 \times 10^6)} (1176) = 186.3 \text{ } \mu s$$

and the maximum justification rate, $f_{j,max}$, is

$$f_{j,max} = \frac{1}{t_j} = 5367 \text{ Hz}$$

The minimum write clock frequency is 1.543799×10^6 Hz. Therefore, the maximum peak value of the waiting time jitter is

$$(1545846.9 - 1543799.0 \text{ bits/sec}) \times 186.3 \text{ } \mu s = 0.38 \text{ bits}$$

The above jitter is relative to the position of the read clock versus the write clock at the instant the justification decision was made.

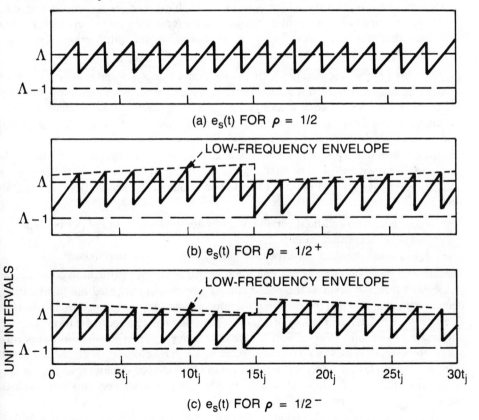

(a) $e_s(t)$ FOR $\rho = 1/2$

(b) $e_s(t)$ FOR $\rho = 1/2^+$

(c) $e_s(t)$ FOR $\rho = 1/2^-$

Fig. 5.12 The effect of waiting time jitter on the synchronizer phase comparator output

The jitter introduced by multiplexes using positive justification has been analyzed by various authors [5.5–5.13]. Both frequency and time domain simulation approaches may be used to examine multiplex jitter generation:

- First order frequency domain analyses that neglect the effect of overhead bits, elastic store size, as well as justification logic timing, have provided expressions for the multiplex jitter power spectral density [5.6, 5.8].
- Theoretical analysis, including the effect of overhead bits, synchronizer elastic store size, and type of phase comparator, showed that in practice there are more jitter components than those predicted by the first order analyses (e.g.,[5.7]). Recent frequency domain analysis has provided a generalized expression for multiplex jitter power spectral density, which includes the effect of the desynchronizer in addition to format and other jitter-related hardware parameters [5.10].
- Time domain analysis, which computes the phase changes associated with data and overhead bits to calculate the jitter waveform for generalized formats, has provided results for rms and peak-to-peak multiplex jitter amplitudes (e.g., [5.13]).

In this section, the first order frequency domain analysis and principles of time domain simulation will be discussed. These analyses characterize multiplex jitter that is strictly a function of format and phase-smoothing circuit transfer characteristic, but may be generalized to allow analysis of jitter accumulation. The results in [5.10] will be useful for optimizing equipment designs.

Two different approaches towards deriving the power spectrum of multiplex jitter are discussed below. The first approach, [5.2, 5.8], decomposes the multiplex jitter waveform,[3] $e_s(t)$, of Figure 5.13a into two components, shown in Figures 5.13b and 5.13c. These components are:

- $e_{s1}(t)$, a sample-and-hold version of a saw-toothed waveform of period $1/\rho f_{j,max}$, and unity amplitude, and
- $e_{s2}(t)$, a saw-toothed waveform of period $1/f_{j,max}$, and amplitude ρ.

This may be understood by examining the scenario in which justification could occur on demand (e.g., as soon as the phase comparator output exceeded the justification decision threshold). Then the resulting waveform would be represented by the dashed line in Figure 5.13a, denoted $e_j(t)$; $e_j(t)$ is a periodic saw-toothed waveform with an amplitude of one unit interval and a period of $1/(\rho f_{j,max})$. At justification opportunities, the sample values of $e_s(t)$ and $e_j(t)$ are equal. Therefore, $e_s(t)$ can be obtained from these samples by applying them to a sample-and-hold circuit, resulting in the waveform of Figure 5.13b, and adding to the result another periodic saw-toothed waveform, $e_{s2}(t)$, as shown in Figure 5.13c.

We can represent $e_j(t)$, shown in Figure 5.14a, by its Fourier series where the dc term is ignored

$$e_j(t) = \frac{1}{\pi} \sum_{n=1}^{\infty} \frac{(-1)^{n+1}}{n} \sin(2\pi n \rho f_{j,max} t) \qquad (5.18)$$

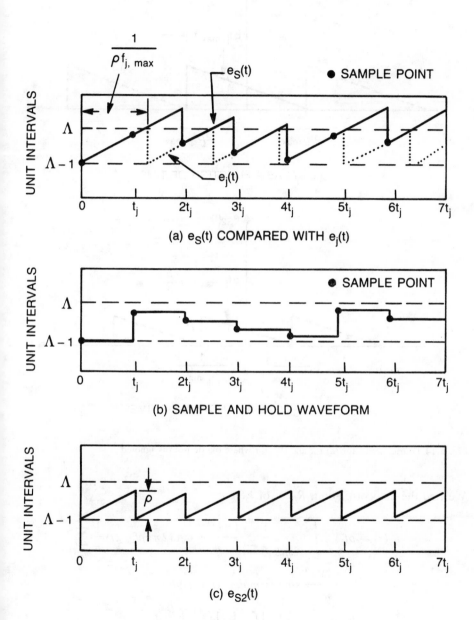

Fig. 5.13 Decomposition of multiplex jitter waveform

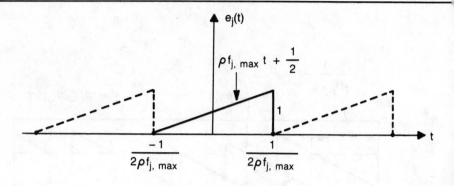

(a) $e_j(t)$ AS A FUNCTION OF TIME

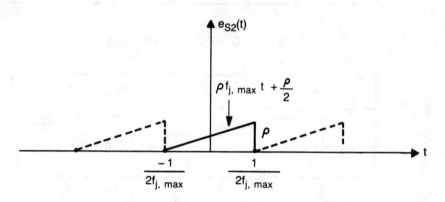

(b) $e_{S2}(t)$ AS A FUNCTION OF TIME

Fig. 5.14 Fourier series for (a) $e_d t$ and (b) $e_{S2} t$ where the dc term is ignored

We find the autocorrelation, $R_j(\tau)$, of $e_j(t)$ as

$$R_j(\tau) = \rho f_{j,max} \int_{-1/2\rho f_{j,max}}^{1/2\rho f_{j,max}} \frac{1}{\pi} \sum_{n=1}^{\infty} \frac{(-1)^{n+1}}{n} \sin\left(2\pi n\rho f_{j,max} t\right) \frac{1}{\pi} \sum_{m=1}^{\infty}$$

$$\cdot \frac{(-1)^{m+1}}{m} \sin\left(2\pi m\rho f_{j,max}(t + \tau)\right) dt$$

$$= \frac{\rho f_{j,max}}{\pi^2} \sum_{n=1}^{\infty} \sum_{m=1}^{\infty} \frac{(-1)^{n+1}}{n} \frac{(-1)^{m+1}}{m} \int_{-1/2\rho f_{j,max}}^{1/2\rho f_{j,max}} \sin\left(2\pi n\rho f_{j,max} t\right)$$

$$\cdot \sin\left(2\pi m\rho f_{j,max}(t + \tau)\right) dt \tag{5.19}$$

Denoting the integral as $I_{nm}(\tau)$, and expanding the term $\sin(2\pi m\rho f_{j,max}(t + \tau))$ we obtain

$$
\begin{aligned}
I_{nm}(\tau) &= \cos(2\pi m\rho f_{j,max}\tau) \int_{-1/2\rho f_{j,max}}^{1/2\rho f_{j,max}} \sin(2\pi n\rho f_{j,max}t) \sin(2\pi m\rho f_{j,max}t)\, dt \\
&\quad + \sin(2\pi m\rho f_{j,max}\tau) \int_{-1/\rho f_{j,max}}^{1/2\rho f_{j,max}} \sin(2\pi n\rho f_{j,max}t) \cos(2\pi m\rho f_{j,max}t)\, dt \\
&= \cos(2\pi m\rho f_{j,max}\tau) \frac{1}{2\pi \rho f_{j,max}} \int_{-\pi}^{\pi} \sin(nt) \sin(mt)\, dt \\
&= \frac{1}{2\pi \rho f_{j,max}} \cos(2\pi m\rho f_{j,max}\tau) \left\{ 2 \cdot \frac{\pi}{2} \cdot \delta_{mn} \right\} \\
&= \frac{\delta_{nm}}{2\rho f_{j,max}} \cos(2\pi m\rho f_{j,max}\tau)
\end{aligned}
\tag{5.20}
$$

Substituting into (5.19), we obtain

$$
\begin{aligned}
R_j(\tau) &= \frac{\rho f_{j,max}}{\pi^2} \sum_{n=1}^{\infty} \sum_{m=1}^{\infty} \frac{(-1)^{n+1}}{n} \frac{(-1)^{m+1}}{m} \frac{\delta_{nm}}{2\rho f_{j,max}} \cos(2\pi m\rho f_{j,max}\tau) \\
&= \frac{1}{2\pi^2} \sum_{n=1}^{\infty} \frac{1}{n^2} \cos(2\pi n\rho f_{j,max}\tau)
\end{aligned}
\tag{5.21}
$$

The power spectrum is obtained by Fourier transforming (5.21), resulting in

$$
\begin{aligned}
\Phi_j(f) &= \int_{-\infty}^{\infty} \sum_{n=1}^{\infty} \frac{1}{2(\pi n)^2} \cos(2\pi n\rho f_{j,max}\tau)\, e^{-j2\pi f\tau}\, d\tau \\
&= \sum_{n=1}^{\infty} \left(\frac{1}{2\pi n} \right)^2 \left\{ \delta(f - n\rho f_{j,max}) + \delta(f + n\rho f_{j,max}) \right\}
\end{aligned}
\tag{5.22}
$$

Sampling $\Phi_j(f)$, at the justification opportunities (illustrated in Figure 5.13a), causes a replication of this spectrum about multiples of the sampling rate to give a spectrum of

$$
\begin{aligned}
Q(f) &= \sum_{n=1}^{\infty} \left(\frac{1}{2\pi n} \right)^2 \sum_{k=-\infty}^{\infty} \left\{ \delta(f - n\rho f_{j,max} - kf_{j,max}) \right. \\
&\quad \left. + \delta(f + n\rho f_{j,max} - kf_{j,max}) \right\}
\end{aligned}
\tag{5.23}
$$

Finally, the hold operation is equivalent to convolution in the time domain with a rectangle function of width equal to one. In the frequency domain, this amounts to multiplication by $\mathrm{sinc}^2(f)$; therefore the spectrum of the first component is given by

$$\Phi_{S1}(f) = \mathrm{sinc}^2(f)Q(f) \tag{5.24}$$

The second component, $e_{S2}(t)$, shown in Figure 5.13b, can be represented by its Fourier series where the dc term is ignored

$$e_{S2}(t) = \frac{\rho}{\pi} \sum_{n=1}^{\infty} \frac{(-1)^{n+1}}{n} \sin(2\pi n f_{j,max} t) \tag{5.25}$$

Taking the Fourier transform of the autocorrelation of $e_{S2}(t)$, the power spectrum of the second component is given by

$$\Phi_{S2}(f) = \sum_{n=1}^{\infty} \left(\frac{\rho}{2\pi n}\right)^2 \{\delta(f - nf_{j,max}) + \delta(f + nf_{j,max})\} \tag{5.26}$$

Thus, the total unfiltered multiplex jitter spectrum is

$$\Phi_{S}(f) = \mathrm{sinc}^2(f)Q(f) + \sum_{n=1}^{\infty} \left(\frac{\rho}{2\pi n}\right)^2 (\delta(f - nf_{j,max}) + \delta(f + nf_{j,max})) \tag{5.27}$$

Note that $\Phi_{S2}(f)$ contains only relatively high frequency components that can generally be easily filtered by the desynchronizer phase-smoothing circuit. The low frequency components of $\Phi_{S}(f)$ are contained in $Q(f)$. If ρ is irrational, none of the spectral lines will coincide. However, when ρ is rational, coincidence occurs and $Q(f)$ may be expressed as a finite sum. If $\rho = p/q$, and p and q are relatively prime, it is shown in [5.14] that

$$Q(f) = \frac{1}{4q^2} \sum_{n=1}^{q-1} \csc^2\left(\frac{n}{q}\pi\right) \sum_{k=-\infty}^{\infty} \delta\left(f - \frac{pn}{q}f_{j,max} - kf_{j,max}\right)$$
$$+ \frac{1}{12q^2} \sum_{k=-\infty}^{\infty} \delta(f - kf_{j,max}) \tag{5.28}$$

If $G(f)$ is the desynchronizer phase-smoothing circuit transfer function, then the filtered demultiplexer jitter output is

$$\Phi_{D}(f) = |G(f)|^2 \Phi_{S}(f)$$

with an rms amplitude of

$$\sigma_D = \sqrt{\int_{-\infty}^{\infty} \Phi_D(f)df} \qquad (5.29)$$

Two examples of experimental measurements of the unfiltered jitter spectrum of a DS1 to DS2 multiplex are shown in Figures 5.15a and 5.15b [5.6]. These figures illustrate $\Phi_S(f)$ for justification ratios of 0.372, and 0.333, respectively. The n^- and n^+ notations on the figures refer to spectral components from the $Q(f)$ term in (5.27), corresponding to spectral lines introduced by

$$\frac{1}{(2\pi n)^2} \sum_{k=-\infty}^{\infty} \delta(f - \rho n f_{j,max} - k f_{j,max})$$

and

$$\frac{1}{(2\pi n)^2} \sum_{k=-\infty}^{\infty} \delta(f + \rho n f_{j,max} - k f_{j,max})$$

respectively. Comparison of Figures 5.15a and 5.15b shows that the spectral lines are all distinct only when ρ is not close to any simple rational number. Note that the vertical scale does not correspond to any particular amount of jitter power. However, the locations and the relative amplitudes of the spectral lines agree very well with the theoretical model.

Figures 5.16a and 5.16b reproduce experimental measurement and theoretical calculation of σ_D as a function of justification ratio for a DS1 to DS2 multiplex [5.6]. The desynchronizer phase-smoothing circuit is a quadratic filter having a damping coefficient of 1.0, and corner frequency of 644 Hz. Again, the dB scale in the figure is only approximate and relative; nevertheless, agreement between experimental and theoretical results is still good. Vertical lines in the tops of the peaks in Figure 5.16a are caused by a low-frequency cutoff in the measurement equipment. Figure 5.16 indicates that the filtered multiplex jitter power is strongly dependent on the choice of ρ. Although it will be shown in Chapter 7 that this dependence is not as significant as might be inferred from the information presented up to this point, justification ratios near 0.0, 0.5, and 1.0 result in maximal jitter generation. It should be emphasized that the spectral components will then be primarily very low frequency and should fall within the desynchronizer phase-smoothing circuit passband.

The second approach, described in [5.6], uses stochastic processes to calculate the power spectrum of multiplex jitter. The basic procedure used is to: (1) find an equation describing multiplex jitter waveforms, (2) introduce initial condition random variables into this equation in such a way that a stationary ensemble of multiplex

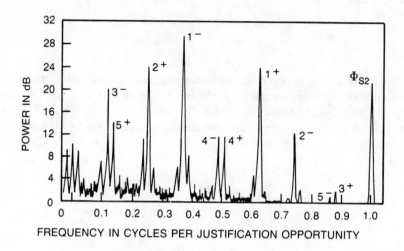

(a) AN EXPERIMENTAL MULTIPLEX JITTER
SPECTRUM WITH $P = 0.372$

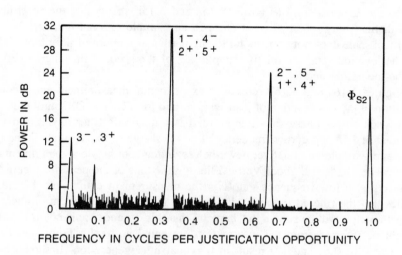

(b) AN EXPERIMENTAL MULTIPLEX JITTER
SPECTRUM WITH $P = 0.333$

Fig. 5.15 Experimental measurements of the unfiltered jitter spectrum

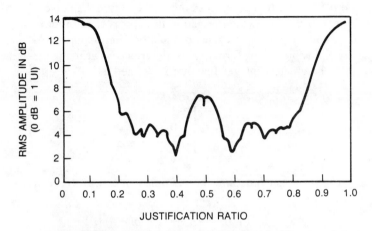

(a) AN EXPERIMENTAL GRAPH OF THE RMS AMPLITUDE OF FILTERED
MULTIPLEX JITTER AS A FUNCTION OF JUSTIFICATION RATIO

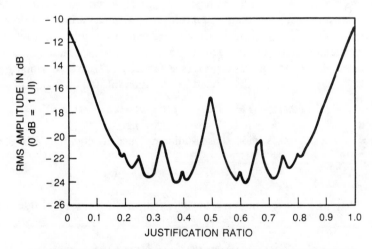

(b) A THEORETICAL GRAPH OF THE RMS AMPLITUDE OF FILTERED
MULTIPLEX JITTER AS A FUNCTION OF JUSTIFICATION RATIO

NOTE: DESYNCHRONIZER FILTER CORNER FREQUENCY = 644 Hz

Fig. 5.16 Experimental measurement and theoretical calculation of σ_D as a function of justification ratio

jitter waveforms is defined, (3) compute the covariance of the multiplex jitter random process, and (4) Fourier transform this covariance to find the power spectrum $\Phi_S(f)$. It is convenient in the following derivation to express time in terms of justification opportunities, and frequency in cycles per justification opportunity.[4]

The multiplex jitter waveform may be described by utilizing the greatest integer function, $[\cdot]$. The greatest integer function is defined as

$$[x] = \begin{cases} \vdots & \vdots \\ -1 & -1 \leq x < 0 \\ 0 & 0 \leq x < 1 \\ 1 & 1 \leq x < 2 \\ \vdots & \vdots \end{cases} \tag{5.30}$$

Using the greatest integer function, the multiplex jitter waveform may be expressed as

$$e_S(t) = (\Lambda - 1) + \rho t - [\rho[t]] \tag{5.31}$$

Referring to Figure 5.13a, the constant term $\Lambda - 1$ causes $e_S(0)$ to be $\Lambda - 1$, the second term ρt generates the linearly increasing portion of the waveform $e_S(t)$, and justifications are made by the third term $[\rho f_{j,max}[t]]$. A stationary ensemble of multiplex jitter waveforms, plotted against a time scale from an arbitrary reference clock is illustrated in Figure 5.17 [5.6]. Defining the random variable τ as how long before $t = 0$ a justification opportunity occurred, and the random variable ξ as how much $e_S(\cdot)$ exceeded $\Lambda - 1$ at time $t = -\tau^+$ (see Figure 5.17), where τ and ξ are independent, and both uniformly distributed on the interval $[0,1)$, $e_S(t)$ is given by

$$e_S(t) = (\Lambda - 1) + \xi + \rho(t + \tau) - [\xi + \rho[t + \tau]] \tag{5.32}$$

The random process $e_S(t)$ is wide sense stationary, and its covariance is given by

$$C_S(t) = E(\{(e_S(t + s) - \mu_S\}\{e_S(s) - \mu_S\}) \tag{5.33}$$

where $\mu_S = E\{e_S(t)\}$ and E denotes expectation. It is shown in [5.6] that

$$C_S(t) = A(t)^* \{w(\rho t) \cdot \sum_{k=-\infty}^{\infty} \delta(t - k)\} + \rho^2 w(t) \tag{5.34}$$

where

$$A(t) = \begin{cases} 1 - |t| & |t| \leq 1 \\ 0 & |t| > 1 \end{cases} \tag{5.35}$$

$$w(t) = \frac{1}{12} - \frac{1}{2} u(t)\{1 - u(t)\} \tag{5.36}$$

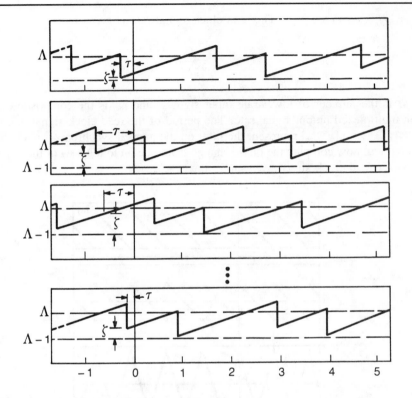

Fig. 5.17 An ensemble of multiplex jitter waveforms

and

$$u(t) = t - [t] \tag{5.37}$$

Graphs of $u(t)$, $v(t)$,[5] and $w(t)$ are given in Figure 5.18 [5.6]. The Fourier transform of (5.33) is [5.6]

$$\Phi_S(f) = \text{sinc}^2(f)Q(f) + \sum_{n=1}^{\infty} \left(\frac{\rho}{2\pi n}\right)^2 \{\delta(f - nf_{j,max}) + \delta(f + nf_{j,max})\} \tag{5.38}$$

where $Q(f)$ is given by (5.23), and frequency and time are once again expressed in traditional units. Thus, the two approaches yield identical results.

Time domain simulations are useful in replicating specific multiplex operation. The multiplex jitter waveform may be calculated by determining the synchronizer phase comparator output. A method for calculating the multiplex jitter waveform using this approach is discussed below [5.11, 5.13].

If the incoming signal is unjittered, then the period of the incoming lower rate signal is $1/f_{nom}$, denoted T_{nom}. The read clock signal rate is given by

$$f_r = \frac{f_{out}}{M} \tag{5.39}$$

where M is the number of lower-rate input signals, and f_{out} is the composite time-division multiplexed output signal rate. The period of the read clock signal is given by $T_r = 1/f_r$. Let $e_S[nT_{nom}]$ represent the jitter on the gapped synchronizer read clock signal. We will now compute the phase change from one bit to the next of the gapped synchronizer read clock signal,[6] $\Delta_1 e_S[nT_{nom}]$.

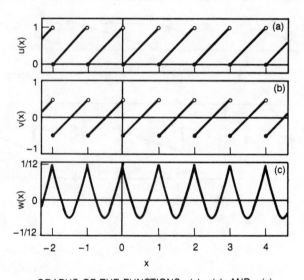

GRAPHS OF THE FUNCTIONS u(x), v(x), AND w(x)

Fig. 5.18 Graphs of the functions $u(x)$, $v(x)$, and $w(x)$

Consider two successive data bits written into the synchronizer elastic store at times nT_{nom} and $(n + 1)T_{nom}$ (for arbitrary n), respectively. If no overhead bits are present from one bit to the next of the synchronizer read clock signal, these bits will be read out of the store at times mT_r and $(m + 1)T_r$ (for arbitrary m), respectively. Let $e_{SPC}[nT_{nom}]$ represent the synchronizer phase comparator output for a bit written into the synchronizer elastic store at time nT_{nom}. Then

$$e_{SPC}[nT_{nom}] = nT_{nom} - mT_r$$

$$e_{SPC}[(n + 1)T_{nom}] = (n + 1)T_{nom} - (m + 1)T_r$$

The phase change from one bit to the next of the synchronizer phase comparator, $\Delta_1 e_{SPC}[nT_{nom}]$, is then

$$\Delta_1 e_{SPC}\,[nT_{nom}] \equiv e_{SPC}[(n+1)T_{nom}] - e_{SPC}[nT_{nom}]$$

$$= (n+1)T_{nom} - (m+1)T_r - (nT_{nom} - mT_r)$$

$$= T_{nom} - T_r$$

If n_{oh} contiguous overhead bits are present from one data bit to the next of the synchronizer read clock signal, then

$$e_{SPC}[(n+1)T_{nom}] = (n+1)T_{nom} - \{(m+1)T_r + n_{oh}T_r\}$$

The phase change from one bit to the next of the gapped synchronizer read clock signal is then

$$\Delta_1 e_S[nT_{nom}] = \Delta_1 e_{SPC}[nT_{nom}]$$

$$= (n+1)T_{nom} - \{(m+1)T_r + n_{oh}T_r\} - (nT_{nom} - mT_r)$$

$$= T_{nom} - (1 + n_{oh})T_r \qquad (5.40)$$

To compute the phase change over a group of m bits (e.g., from one justification signaling bit to the next) for the gapped synchronizer read clock signal, the phase change between each pair of consecutive bits in the synchronizer read clock signal may be summed. The effect of justification must also be included: specifically, whether the justification opportunity in the format contains a data bit or a justification bit. Defining N_{data} and N_{oh} as the number of data and overhead bits for the designated group of bits, respectively, and assuming that a justification opportunity falls within this group of bits, it is clear from (5.40) that $\Delta_m e_S[nT_{nom}]$ for a group of m bits is

$$\Delta_m e_S[nT_{nom}] = \Delta_m e_{SPC}[nT_{nom}]$$

$$= N_{data}(T_{nom} - T_r) - N_{oh}T_r + \begin{cases} T_{nom} - T_r & \text{no justification} \\ -T_r & \text{justification} \end{cases} \qquad (5.41)$$

Using the techniques discussed above to simulate the synchronizer mechanism, an appropriate desynchronizer filter transfer characteristic should be chosen so that the filtered multiplex output jitter may be determined.

The remainder of this section briefly outlines negative, positive-negative, and positive-zero-negative justification schemes, and references techniques for reducing justification induced jitter.

• Negative justification is the complement of positive justification. Negative justification may also be described making use of the commutator analogy illustrated in Figure 5.7. In this case, the input signal data rate is faster than the common output data rate; therefore, the commutator write brush begins to catch up to the commutator read brush. If the write brush were allowed to overtake the read brush, data would be lost, as it would be overwritten before it has been read (a "slip" will have occurred). To avoid this, the phase comparator monitors the phase difference between the read and write brushes and when a slip appears imminent, an extra bit is read out. Implementation of the negative bit justification process involves assigning, within the multiplexer format, bit positions that are used for signaling a negative justification decision, and negative justification opportunities. The jitter generated for negative justification may be computed analogously to that for positive justification, and the results are identical.

- In the case of positive-negative and positive-zero-negative justification the nominal justification ratio is intended to be zero.[7] Specifically the rate of the synchronized lower-rate signals is nominally the same as the incoming signal rate. Therefore, the synchronized bit rate may be slightly higher, the same, or lower than that of a lower-rate input. If the synchronized bit rate is slightly higher than that of the lower-rate input signal, occasional positive justification will occur; if it is slightly lower, occasional negative justification will occur. If the synchronized bit rate is identical to that of the lower-rate input signal, no justification will be performed: the justification ratio is zero and no multiplex jitter arising from justification will be generated. From a practical standpoint, the justification ratio will typically be *close to,* but not actually, zero. However, at values close to zero the rms jitter amplitude achieves its maximum. Analysis of positive-negative and positive-zero-negative justification has been treated by various authors [5.15–5.17]. Further study of these justification schemes under various initial conditions is required.

- Techniques for reducing jitter introduced by justification have also been investigated [5.18–5.20]. For the case of positive (or negative) justification, methods have been considered to increase the usable range of justification ratios, such as cases in which a format which may have been optimized for a particular application requires usage of an undesirable justification ratio. Techniques for jitter suppression for positive-negative justification have also been explored [5.16].

5.3 MULTIPLEXING USING POINTER PROCESSING

An entire new family of rates and formats for fiber optic transmission systems has been defined in ANSI standard T1.105-1988 [5.21]. The standard defines a basic signal rate and format with a byte (eight bit) interleaved multiplexing scheme to establish an entire family of rates and formats for optical interfaces. The new basic signal rate is at 51.84 Mb/s and the family of rates and formats is defined at a rate of $N \times 51.84$ Mb/s, where N is an integer. Thus, there is an integer multiple relationship between the basic signal rate and the multiplexed signal rates. The basic rate signal is denoted *synchronous transport signal level-1* (STS-1), and an integer multiple, N, of this signal is denoted an STS-N signal. The basic signal can be divided into a portion assigned for overhead needed to form an STS-N signal and a portion which contains the payload, denoted the STS-1 *synchronous payload envelope* (SPE). A network employing this hierarchy is called a *synchronous optical network* (SONET).

The synchronization mechanism used in synchronous optical network equipment is called pointer processing. Recalling the discussion of Section 5.1, the frame format of a multiplexer using slip buffering synchronization allows direct identification of individual bit positions once framing occurs. The SONET multiplex frame format similarly allows for identification of individual bit positions, with an additional level of indirect addressing. Specifically, the number of byte positions between the framing byte and any particular kth data byte is not a constant, but may be determined indirectly through the pointer. This is illustrated in Figure 5.19. Figure 5.19a illustrates the fixed relationship between the framing bit position and any particular kth data bit position in the frame format of a slip buffering synchronization multiplexer. Figure 5.19b illustrates the pointer mechanism. The pointer, P, performs both the functions of framing and synchronization. The number of byte positions between the framing byte, F, and P is a constant, i. P indicates a value which represents the number of byte positions, $p(t)$, to the location of a reference byte position in the format (the beginning of the SPE). This value can be incremented or decremented once every four frames. From this reference byte position, the number of byte positions to any particular kth byte position is fixed. (Note that incrementing-maintaining-decrementing the pointer "value" allows byte adjustments to be made that accommodate frequency and phase variations on the lower-rate input signals, and is conceptually similar to positive-negative justification.)

A block diagram of a pointer processor for a SONET multiplex (or other synchronous optical network element) with an STS-N input is contained in Figure 5.20. The pointer processor operation is actually analogous to that of a desynchronizer-resynchronizer. In the processor the received STS-N clock is extracted from the incoming signal, divided by N, and used to clock the STS-1 signals into associated pointer processors. Within a pointer processor this clock is inhibited to account for incoming (positive) pointer adjustments, as well as for STS-1 frame overhead byte occurrences (not data arising from negative pointer adjustments) so that only data bytes are written into the byte-oriented elastic store. The phase comparator output indicates the phase difference between the read and write clocks, which is monitored by control circuitry. Whenever the phase comparator output indicates the store is becoming filled or depleted (upper or lower pointer adjustment decision threshold has been crossed, respectively), a decision is made to perform an outgoing negative or positive pointer adjustment at the next pointer adjustment opportunity. The data bytes (comprising the STS-1 SPE) are then read out of the byte oriented elastic store

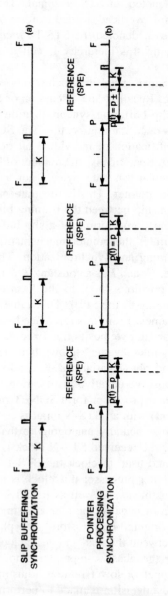

Fig. 5.19 Multiplexing format comparison

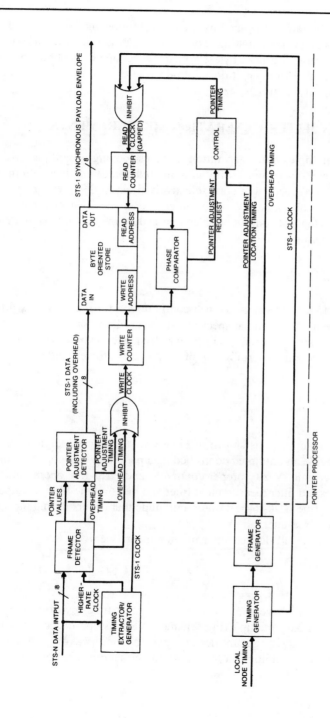

Fig. 5.20 SONET multiplex pointer processor block diagram

at a STS-1 read clock rate determined by local synchronous node timing, which has been inhibited to account for outgoing (positive) pointer adjustments, as well as for insertion of STS-1 frame overhead bytes. Pointer processing does not introduce jitter onto the STS-N signal. However, it does introduce jitter onto the payload it carries. This will be discussed within Chapter 7.

5.4 MULTIPLEX JITTER GENERATION MEASUREMENTS

Measurement of multiplex jitter generation involves looping a multiplexer-demultiplexer pair at the high-speed ports with an unjittered controlled data pattern applied at the multiplexer low-speed input port, selectively filtering the jitter, and measuring the rms or peak-to-peak amplitude of the jitter over a specified measurement time interval. The following test configuration and procedure may be utilized to determine multiplex jitter generation [5.22].

Test Configuration

Figure 5.21 [5.22] illustrates the test configuration for the multiplex jitter generation measurement. The optional spectrum analyzer allows observation of the output jitter frequency spectrum, and the optional frequency synthesizer may be used to provide a more accurate determination of frequencies utilized in the measurement procedure.

Procedure

1. Connect the equipment as shown in Figure 5.21, using the digital signal generator to provide an unjittered controlled data pattern to the looped multiplexer-demultiplexer pair. Verify proper continuity and error-free operation.
2. Select the desired jitter measurement filter and measure the filtered output jitter, recording the true peak-to-peak jitter amplitude that occurs during the specified measurement time interval.
3. Repeat Step 2 for all desired jitter measurement filters.

5.5 REMARKS

Within this chapter, we have examined the concept of time-division multiplexing, and discussed the primary multiplexing schemes. Various approaches for characterizing multiplex jitter power spectral density, and rms and peak-to-peak amplitudes have been presented. In addition, a technique for measuring multiplex jitter generation has been described.

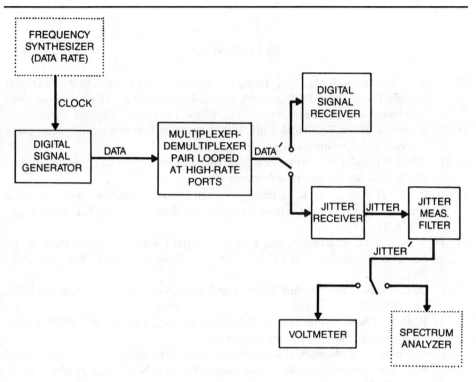

Fig. 5.21 Multiplex jitter measurement configuration

NOTES

1. Some clocks, upon experiencing short interruptions of the input timing reference, do not operate in strict phase-lock with it.
2. In the absence of input jitter, the incoming and nominal line rates are equal.
3. The multiplex jitter waveform is often denoted as the waiting time jitter waveform, even though it is understood that this waveform is actually the sum of waiting time jitter and the jitter which would always be present due to justification.
4. Conversion to seconds and Hertz are through the relations: (1) 1 justification opportunity = t_j, and (2) 1 cycle per justification opportunity = $f_{j,max}$
5. $v(t) = u(t) - 1/2$
6. Using (7.60) it may be shown that in the absence of input jitter the phase change from one bit to the next of the gapped synchronizer read clock signal is the same as for the synchronizer phase comparator output.
7. Positive-zero-negative justification differs from positive-negative justification in that consecutive transitions of bit justifications of different polarity are not allowed.

REFERENCES

[5.1] V. I. Johannes, R. H. McCullough, "Multiplexing of Asynchronous Digital Signals Using Pulse Stuffing with Added-Bit Signaling," *IEEE Transactions on Communication Technology,* Vol. COM-14, No. 5, October 1966.

[5.2] Members of the Technical Staff, Bell Telephone Laboratories, *Transmission Systems for Communications,* Fifth Edition, 1982.

[5.3] ANSI T1.102-1987, *American National Standard for Telecommunications-Digital Hierarchy-Electrical Interfaces,.*

[5.4] C. E. Huffman, J. K. Blake, "Asynchronous Multiplex Jitter," *Collins Transmission Systems Division Technical Bulletin,* 523-0605721-00283J, pp. 6-1–6-18.

[5.5] Y. Matsuura, S. Kozuka, and K. Yuki, "Jitter Characteristics of Pulse Stuffing Synchronization," *IEEE Int. Conf. on Commun.,* June 1968, pp. 259–264.

[5.6] D. L. Duttweiler, "Waiting Time Jitter," *BSTJ* Vol. 51, No. 1, January 1972, pp. 165–207.

[5.7] P. E. K. Chow, "Jitter Due to Pulse Stuffing Synchronization," *IEEE Trans. on Commun.,* July 1973, pp. 854–859.

[5.8] F. Kuhne, "Mathematical Representation of Justification Jitter," *Arch. Elektron. and Uebertragungstech.* (Germany), Vol. 33, No. 7–8, July–Aug. 1979, pp. 327–329.

[5.9] F. Kuhne, N. Kamp, H. Muller, "Intrinsic Jitter of Digital Multplex Equipment," *Arch. Elektron. and Uebertragungstech,* (Germany), Vol. 33, No. 11, Nov. 1979, pp. 462–464.

[5.10] P. E. K. Chow, "Generalized Formula for Pulse Stuffing Jitter," *Electronics Letters,* Vol. 22, No. 24, Nov. 20, 1986, pp. 1313–1314.

[5.11] H. J. Altman, A. G. Kaim, T1X1.3 Jitter and Wander Subworking Group Contribution JWC/86-042, *Multiplexer Jitter Performance: An Analysis of the Stuffing Mechanism,* T1 Committee, 1986.

[5.12] J. A. Gracia, R. H. Huhn, *Using Simulation to Analyze Pulse Stuffing Network Jitter,* Winter Simulation Conference, Dec. 5–7, 1977.

[5.13] R. O. Nunn, *Unpublished work on A Time-Domain Simulator for Waiting Time Jitter/Wander,* AT&T Bell Laboratories, 1987.

[5.14] J. E. Iwerson, "Calculated Quantizing Noise of Single-Integration Delta-Modulation Coders," *BSTJ,* 48, No. 7, part 3, September 1969, pp. 2359–2389.

[5.15] "Improved Method of Justification," Annex 1 to Recommendation G.741, *CCITT Green Book,* Volume III.

[5.16] K. Inagaki, Y. Hirata, A. Ogawa, "A New Technique for Data Transmission Via TDMA Satellite Link," *NTC* 77.

5.17] F. Kuhne, K. Lang, "A Positive-Zero-Negative Stuffing Technique For the Multiplex Transmission of Plesiochronous Data Signals," *FREQUENZ* 329109:281-17, 1978.

5.18] J. D. Heightly, V. I. Johannes, J. S. Mayo, F. J. Witt, *Jitter Reduction in Pulse Multiplexing Systems Employing Pulse Stuffing,* United States Patent Office, Patent 3,420,956, January 7, 1969.

5.19] W. D. Grover, T. E. Moore, J. A. McEachern, "Measured Pulse-Stuffing Jitter in Asynchronous DS-1/SONET Multiplexing With and Without Stuff-Threshold Modulation Circuit," *Electronics Letters,* Vol. 23, No. 18, August 27, 1987, pp 959–961.

5.20] W. D. Grover, T. E. Moore, J. A. McEachern, "Waiting Time Jitter Reduction by Synchronizer Stuff-Threshold Modulation," *GLOBECOM '87 Proceedings,* Tokyo.

5.21] ANSI T1.105-1988, "Digital Hierarchy Optical Interface Rates and Formats Specification."

5.22] T1X1.4/87-024, "Jitter Measurement Methodology," T1 Committee, 1987.

5.23] ANSI T1.101-1987, "Synchronizatiton Interface Standards For Digital Networks."

Chapter 6
Jitter Tolerance and Transfer
in Digital Multiplexes

The tolerance of a digital multiplex to input jitter is its ability to perform satisfactorily when the lower-rate signals input to the multiplexing terminal, or the higher-rate signal input to the demultiplexing terminal is jittered. Inadequate tolerance to input jitter may result in excessive synchronizer or desynchronizer elastic store slips, and bit errors. Digital multiplex tolerance to input jitter is primarily determined by factors relating to input timing extraction circuits, synchronizer and desynchronizer elastic store size, or justification or pointer adjustment capacity.

The jitter transfer of an individual digital equipment is defined as the ratio of the output jitter to the applied input jitter as a function of frequency. Jitter transfer for line regenerators has been described in Chapter 2. Jitter transfer for multiplexes using justification techniques is more complex because of the non-linearity of the multiplexer justification mechanism. The concept of jitter transfer is not applicable for multiplexes using slip buffering or pointer processing, since these mechanisms do not transfer input jitter to the multiplex output. In this chapter, we will study the jitter tolerance and jitter transfer of digital multiplexes.

6.1 MULTIPLEX JITTER TOLERANCE AND MEASUREMENTS

The tolerance of a digital multiplex to input jitter may be categorized as a function of the synchronization scheme it utilizes to attain bit rate equalization. Although some treatment of the jitter tolerance of multiplexes using slip buffering and pointer processing synchronization will be included in this section, the primary emphasis will be on the tolerance of digital multiplexes using positive bit justification. The techniques used to analyze positive bit justification multiplex tolerance are open to generalization and provide insight into pointer adjustment multiplex tolerance as well.

As discussed in Chapter 5, the multiplexer frame format determines the maximum bit justification rate, and can therefore only accommodate limited positive and negative frequency deviations from the nominal incoming line frequency. Input jitter can cause positive and negative frequency deviations which can saturate the justification process. The point at which the justification process can no longer accommodate all of the input jitter is defined as the onset of saturation. The remainder of the input jitter must be accommodated in the synchronizer store. The applied input jitter which is encoded by the justification mechanism is reproduced exactly at the input to the desynchronizer store.

The following discussion defines and provides insight into the four regions of multiplex jitter tolerance [6.1]. As an example, Figure 6.1 [6.1] illustrates these four regions of tolerance for applied sinusoidal input jitter.

- Within Region 1, the multiplex performance is limited by the capacity of the desynchronizer elastic store, and by the degree to which the phase-smoothing circuit follows jitter and passes it on to the output. Within this region, the frequency deviation from the nominal of the applied sinusoidal jitter is below that required for onset of saturation of the synchronizer store; the synchronizer store is not absorbing any of the applied jitter, and the only jitter within it exists because of justification and waiting time. Thus, as the input jitter amplitude is increased in this region, the jitter applied to the desynchronizer store increases, and, to the extent that the phase-smoothing circuit removes this jitter, it increases in the desynchronizer store. When a jitter peak of sufficient magnitude coincides with the normal justification and waiting time jitter, a slip occurs, and data errors result. Jitter tolerance in this region decreases with increasing jitter frequency because the phase-smoothing circuit is less able to track high frequency jitter, and thus increases the amount of jitter which must be absorbed by the desynchronizer store.
- Within Region 2, the instantaneous frequency deviation from the nominal caused by the applied sinusoidal jitter cannot be accommodated in the justification range defined by the higher level multiplex format. Some of the applied jitter is absorbed in the synchronizer store, and the total applied jitter which can be tolerated before the desynchronizer store slips increases. Jitter tolerance in this region increases with an increase in frequency of the applied jitter, because more of the applied jitter is absorbed by the synchronizer store. Eventually, as the frequency of the input jitter continues to increase, the synchronizer store capacity will be exceeded on a + or − peak of applied jitter before the desynchronizer elastic store slips. This marks the beginning of Region 3.
- Within Region 3, the tolerance of the multiplex to the applied sinusoidal jitter is limited by the capacity of the synchronizer store. The desynchronizer store will no longer slip as the applied jitter frequency increases above this point, since the peak jitter encoded in the data stream is decreasing with increasing jitter frequency. The synchronizer store and justification format act as a limiter in this region. Jitter tolerance decreases with increasing frequency in this region, because the synchronizer store capacity is fixed, and the portion of the

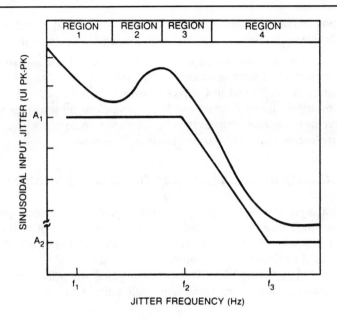

Fig. 6.1 Four regions of multiplex jitter tolerance

input jitter cycle which is encoded in the high level multiplex data stream is decreasing. When the frequency of the applied jitter begins to exceed the bandwidth of the input clock extraction circuit, the extracted clock will no longer track the input data. As the mistracking becomes larger, errors will occur at the data detection point, where the input data is retimed to the extracted clock; this marks the transition to Region 4.

- Within Region 4, the input jitter tolerance will continue to decrease with increasing frequency of applied jitter, due to data detection errors, until the jitter amplitude falls within the width of the "eye" allocated to input jitter (see Chapter 4). Input jitter below this amplitude can be accepted error-free at any frequency.

Several approaches may be used to determine the relationship between multiplex input jitter tolerance, bit justification capacity, and synchronizer and desynchronizer store sizes:

- Analysis of bit justification capability to track input jitter.
- Estimation of worst-case bounds on the synchronizer and desynchronizer store sizes for sinusoidal input jitter.
- Time domain approaches, using sinusoidal input jitter, that contribute to an intuitive understanding of phase accumulation within synchronizer and desynchronizer stores.
- Time domain simulation analyses that compute multiplex jitter tolerance as a

function of multiplex frame format, justification strategy, synchronizer and de-synchronizer store sizes, and phase-smoothing circuit implementation.

Within this section, the first three approaches are reviewed and discussed. However, we will not treat the fourth approach within this section. The principles of time domain simulation, described in Chapter 5, may be generalized to study the impact of various multiplex frame formats and implementations on tolerance to input jitter. Relevant events in the elastic stores may be simulated, including read and write clock phases, justification requests, and justification occurrences.

6.1.1 Relationship Between Input Jitter Tolerance and Multiplex Parameters

In order to analyze the capability of the positive justification mechanism to accommodate input jitter, we will make use of the basic principles of positive bit justification. The following discussion ignores, for simplicity, the effect of the overhead bits. As defined in Chapter 5, the nominal justification rate $f_{j,nom}$, is the difference frequency $\bar{f_r} - f_{nom}$. Normal justification takes place when the *actual* justification rate is less than or equal to the maximum justification rate.

$$0 \le \bar{f_r} - f(t) \le f_{j,max} \qquad (6.1)$$

where the incoming line frequency is denoted simply as $f(t)$.

When input jitter, $e_I(t)$, is present, the instantaneous incoming line frequency is deviated from the nominal frequency, f_{nom}. Denoting $\dot{e}_I(t)$ as the frequency deviation caused by the input jitter,

$$f(t) = f_{nom} + \dot{e}_I(t) \qquad (6.2)$$

Thus, the requirement for normal justification is

$$0 \le \bar{f_r} - (f_{nom} + \dot{e}_I(t)) \le f_{j,max} \qquad (6.3)$$

Regrouping the terms, and recalling the definition of $f_{j,nom}$

$$f_{j,nom} - f_{j,max} \le \dot{e}_I(t) \le f_{j,nom} \qquad (6.4)$$

Therefore, the justification control logic can track jitter changes limited by the nominal and maximum justification rates. The capability of the justification mechanism to track input jitter may be examined for the cases of both deterministic and Gaussian input jitter.

If the incoming jitter causes positive frequency deviations from the nominal, the maximum tolerable positive frequency deviation from the nominal which satisfies (6.3) is

$$\dot{e}_I(t) = \bar{f}_r - f_{nom} = f_{j,nom} \equiv f_P \tag{6.5}$$

Based on allowable line frequency tolerances, assign the lowest value f_P can assume as f_{P-}, and the highest value as f_{P+}. Note that a lower value of f_P implies earlier saturation of the justification mechanism.

Similarly, if the incoming jitter causes negative frequency deviations from the nominal, the magnitude of the maximum tolerable negative frequency deviation from the nominal which satisfies (6.3) may be determined from

$$\dot{e}_I(t) = -f_{j,max} + \bar{f}_r - f_{nom} = -(f_{j,max} - f_{j,nom}) \equiv -f_N \tag{6.6}$$

Comparison of the defining expressions for f_P and f_N indicates that $f_N = f_{j,max} - f_P$. Again, based on allowable line frequency tolerances, assign the lowest magnitude f_N can assume as f_{N-} and the highest magnitude as f_{N+}. Note that a lower magnitude of f_N implies earlier saturation of the justification mechanism.

Thus, as long as

$$-f_N \leq \dot{e}_I(t) \leq f_P \tag{6.7}$$

or more conservatively

$$|\dot{e}_I(t)| \leq \min(f_P, f_N)$$

the justification mechanism will accommodate the input jitter. For deterministic input jitter, this is practical to assess.

For Gaussian input jitter, the mean square value of $\dot{e}_I(t)$, $\dot{\sigma}_I^2$, is given by

$$\dot{\sigma}_I^2 = 4\pi^2 \int_{-\infty}^{\infty} f^2 \Phi_I(f) df \tag{6.8}$$

where $\Phi_I(f)$ denotes the power spectrum of the input jitter. Since the input jitter has a Gaussian amplitude distribution, its derivative will have a Gaussian amplitude distribution, since the effect of any linear signal process is to leave a Gaussian process Gaussian. Therefore, as a rough approximation, if

$$6\dot{\sigma}_I \leq \min(f_P, f_N) \tag{6.9}$$

the bit justification mechanism will absorb the bulk of the input jitter [6.2].

Let us now consider the case in which the justification mechanism is unable to absorb most of the input jitter. Whether or not the elastic stores spill (or are depleted) is a function of the amplitude and time duration of the frequency deviation, the maximum justification rate, the justification ratio, and the elastic store sizes. We will first examine the case of a constant frequency deviation, which we will denote by f_I. Consider positive frequency deviations, of magnitude f_I, from the nominal incoming line frequency, such that

$$f_I > f_P$$

Defining M_{S+} as the number of cells in the synchronizer elastic store needed to accommodate positive deviations from f_{nom} and T_+ as the time duration of the positive deviations,

$$\frac{T_+}{M_{S+}} \leq \frac{1}{f_I - f_P} \qquad (6.10)$$

Now consider negative frequency deviations from the nominal incoming line frequency, of magnitude f_I, such that

$$f_I > f_N$$

Defining M_{S-} as the number of cells in the synchronizer elastic store needed to accommodate negative deviations from the nominal incoming line frequency, and T_- as the time duration of the negative deviations,

$$\frac{T_-}{M_{S-}} = \frac{1}{f_I - f_N} \qquad (6.11)$$

Defining M_S as the total number of cells in the synchronizer elastic store needed to accommodate both positive and negative frequency deviations from the nominal ($M_S = M_{S+} + M_{S-}$), and T as the time duration of the positive or negative frequency deviation for *symmetrical jitter* ($T = T_- = T_+$), we may complete the derivation of the equation which relates synchronizer elastic store size, maximum justification rate and justification ratio to the time duration and amplitude of the frequency deviation caused by jitter.

If $\rho < .5$, then there is greater accommodation of negative than positive incoming frequency deviations and $f_P < f_N$. In this case,

$$f_I < f_P: M_S = 0 \qquad (6.12)$$

$$f_P \leq f_I < f_N: M_S = T\{f_I - f_P\} \qquad (6.13)$$

$$f_I \geq f_N: M_S = 2T\{f_I - (f_P + f_N)/2\} \qquad (6.14)$$

The applied input jitter which is encoded into the higher-rate signal justification locations is carried through the multiplex system and reproduced exactly at the desynchronizer elastic store input.

We will now examine the case of frequency deviations caused by sinusoidal input jitter. Sinusoidal input jitter may be defined as

$$e_I(t) = A_I \sin (2\pi f_I t) \qquad (6.15)$$

where A_I is the peak amplitude in terms of unit intervals. It causes frequency deviations given by

$$\dot{e}_I(t) = A_I 2\pi f_I \cos (2\pi f_I t) \qquad (6.16)$$

The magnitude of the peak frequency deviation from the nominal is then given by

$$2\pi A_I f_I \qquad (6.17)$$

We may now estimate a very loose bound on the synchronizer and desynchronizer store sizes. The number of bits written into the synchronizer elastic store in t seconds is given by the integral of (6.16) over the time interval t. Therefore, during the positive half cycle of the input jitter an extra $2A_I$ data bits are written into the synchronizer elastic store compared to the unjittered nominal number, and during the negative half cycle $2A_I$ fewer are written. If the effect of the justification ratio were not considered, it would be necessary for the elastic store size to be at least $4A_I$ bits just for input jitter (assuming the clock spacing was half the store size at the beginning of the input jitter cycle).

If during the positive half cycles of the input jitter none of the justification opportunities were used, then

$$\bar{f}_r - (f_{nom} + \dot{e}_I(t)) = 0 \qquad (6.18)$$

or

$$\dot{e}_I(t) = \bar{f}_r - f_{nom} = \rho f_{j,max} \qquad (6.19)$$

Thus, for half a cycle $1/2f_I$, a total of $\rho f_{j,max}/2f_I$ of these $2A_I$ extra data bits could be transmitted. If during the negative half cycles of the input jitter all of the justification opportunities were used, then

$$\bar{f}_r - (f_{nom} + \dot{e}_I(t)) = f_{j,max} \qquad (6.20)$$

$$\dot{e}_I(t) = -(1 - \rho)f_{j,max} \qquad (6.21)$$

Thus, for half a cycle $1/2f_I$, a total of $(1 - \rho)f_{j,max}/2f_I$ fewer data bits could be transmitted.

Therefore, a very loose bound on the synchronizer elastic store size, M_S, may be given by

$$M_S > 2\left(2A_I - x\frac{f_{j,max}}{2f_I}\right)$$ (6.22)

where

$$x = \min\left[\rho, (1 - \rho)\right]$$

The jitter which is encoded in the justification mechanism is recovered, and must be accommodated, in the desynchronizer elastic store. We will now estimate a bound on the desynchronizer elastic store size. During the positive half cycles of the input jitter, $\rho f_{j,max}/2f_I$ more data bits were transmitted to the desynchronizer. During the negative half cycles of the input jitter, $(1 - \rho)f_{j,max}/2f_I$ more justification bits were transmitted. Therefore, a loose bound on the desynchronizer elastic store for this case is

$$M_D > 2x\frac{f_{j,max}}{2f_I} = x\frac{f_{j,max}}{f_I}$$ (6.23)

where

$$x = \max\left[\rho, (1 - \rho)\right]$$

We will now consider a time domain approach [6.1] which contributes to an intuitive understanding of phase accumulation in synchronizer and desynchronizer elastic stores, while providing a tighter bound on store sizes.

Figures 6.2a and 6.2b illustrate input sinusoidal and maximum tolerable positive and negative frequency deviation variables for the initial conditions:

$$e_I(t = 0) = 0$$ (6.24)

$$\dot{e}_I(t = 0) = 2\pi A_I f_I$$ (6.25)

As illustrated in Figure 6.2b, the magnitude of $\dot{e}_I(t) > f_P$ from time $t = 0$ until $t = t_{P,1}$, where $t_{P,1}$ is defined by

$$\dot{e}_I(t_{P,1}) = f_P = 2\pi A_I f_I \cos(2\pi f_I t_{P,1})$$ (6.26)

or

$$t_{P,1} = \frac{1}{2\pi f_I}\cos^{-1}\left(\frac{f_P}{2\pi A_I f_I}\right)$$ (6.27)

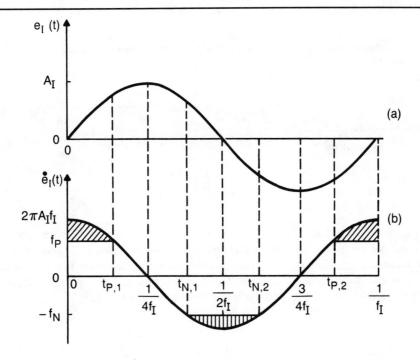

SINUSOIDAL JITTER, e_I (t) = $A_I \sin 2\pi f_I t$, ILLUSTRATED FOR
INITIAL CONDITIONS: e_I (t = 0) = 0, $\overset{\bullet}{e}_I$ (t = 0) = $A\pi$ $A_I f_I$

NOTES: f_P - MAXIMUM TOLERABLE POSITIVE FREQUENCY DEVIATION
FROM NOMINAL

f_N - MAGNITUDE OF MAXIMUM TOLERABLE NEGATIVE FREQUENCY
DEVIATION FROM NOMINAL

Fig. 6.2 Frequency deviation variables for sinusoidal jitter, $e_I = A_I \sin 2\pi f_I t$, illustrated for initial conditions: $e_I(t = 0) = 0$, $e_I(t = 0) = A \pi A_I f_I$

and from time $t = t_{P,2}$ until $t = 1/f_I$ where $t_{P,2}$ is defined by

$$t_{P,2} = \frac{1}{f_I} - t_{P,1} \tag{6.28}$$

or

$$t_{P,2} = \frac{1}{f_I} - \frac{1}{2\pi f_I} \cos^{-1} \frac{f_P}{2\pi A_I f_I} \tag{6.29}$$

In addition, the magnitude of $\dot{e}_I(t) > f_N$ from the time $t_{N,1}$ until $t_{N,2}$, where $t_{N,1}$ is defined by

$$t_{N,1} = \frac{1}{2\pi f_I} \left\{ \pi - \cos^{-1} \frac{f_N}{2\pi A_I f_I} \right\} \tag{6.30}$$

and $t_{N,2}$ is defined by

$$t_{N,2} = \frac{1}{f_I} - t_{N,1} \tag{6.31}$$

Consider the effect of subjecting at $t = 0$ a synchronizer elastic store, operating at its normal fill point, to input sinusoidal jitter as illustrated in Figures 6.2a and 6.2b. If $2\pi A_I f_I > f_P$, (6.3) cannot remain satisfied, as the maximum tolerable positive frequency deviation from the nominal has been exceeded. Thus, positive phase will begin to accumulate in the synchronizer elastic store at the rate

$$f_{net}(t) = f(t) - \bar{f}_r = f_{nom} + \dot{e}_I(t) - \bar{f}_r$$
$$= \dot{e}_I(t) - f_P \tag{6.32}$$

$f_{net}(t)$ will be continuously positive while $\dot{e}_I(t) > f_P$ over the interval $-t_{P,1} \le t < t_{P,1}$ (see Figure 6.2b). Referring to Figure 6.3, the maximum positive excursion of the synchronizer elastic store during this interval, M_{S+}, is given by

$$M_{S+} = 2 \int_0^{t_{P,1}} f_{net} dt = 2 \int_0^{t_{P,1}} \{\dot{e}_I(t) - f_P\} dt$$
$$= 2A_I \sqrt{1 - \left(\frac{f_P}{2\pi A_I f_I}\right)^2} - \frac{1}{\pi} \left(\frac{f_P}{f_I}\right) \cos^{-1} \left(\frac{f_P}{2\pi A_I f_I}\right) \tag{6.33}$$

Similarly, when the magnitude of the negative deviation becomes greater than f_N, negative phase will begin to accumulate in the synchronizer elastic store at the rate

$$f_{net}(t) = f(t) + f_{j,max} - \bar{f}_r = f_{nom} + \dot{e}_I(t) + f_{j,max} - \bar{f}_r$$
$$= \dot{e}_I(t) - (-f_N) \tag{6.34}$$

The maximum negative excursion of the synchronizer elastic store, M_{S-}, during this interval is given by

$$M_{S-} = \int_{t_{N,1}}^{t_{N,2}} f_{net}dt = \int_{t_{N,1}}^{t_{N,2}} \{\dot{e}_I(t) - (-f_N)\}dt$$

$$= -2A_I \sqrt{1 - \left(\frac{f_N}{2\pi A_I f_I}\right)^2} + \frac{1}{\pi}\left(\frac{f_N}{f_I}\right)\cos^{-1}\left(\frac{f_N}{2\pi A_I f_I}\right) \quad (6.35)$$

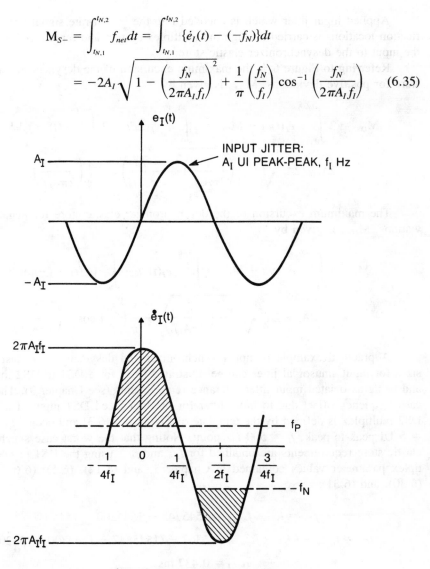

EFFECT OF INPUT JITTER ON SYNCHRONIZER
ELASTIC STORE

NOTE: THE MAXIMUM SYNCHRONIZER STORE REQUIREMENTS
ARE OBTAINED FOR f_{P-} AND f_{N-}

Fig. 6.3 Effect of input jitter on synchronizer elastic store

Applied input jitter which is encoded into the higher-rate signal stream justification locations is carried through the multiplex system and reproduced exactly at the input to the desynchronizer elastic store.

Referring to Figure 6.4, the maximum excursion of the desynchronizer elastic store for positive deviations, M_{D+}, is given by

$$M_{D+} = 2 \int_0^{1/4f_I} \dot{e}_I(t)dt - M_{S+} = 2 \int_0^{1/4f_I} \dot{e}_I(t)dt - 2 \int_0^{t_{P,1}} \{\dot{e}_I(t) - f_P\}dt$$

$$= 2A_I - 2A_I \sqrt{1 - \left(\frac{f_P}{2\pi A_I f_I}\right)^2} + \frac{1}{\pi}\left(\frac{f_P}{f_I}\right)\cos^{-1}\left(\frac{f_P}{2\pi A_I f_I}\right) \tag{6.36}$$

The maximum excursion of the desynchronizer elastic store for negative deviations, M_{D-}, is given by

$$M_{D-} = \int_{1/4f_I}^{3/4f_I} \dot{e}_I(t)dt - M_{S-} = \int_{1/4f_I}^{3/4f_I} \dot{e}_I(t)dt - \int_{t_{N,1}}^{t_{N,2}} \{\dot{e}_I(t) - (-f_N)\}dt$$

$$= -2A_I + 2A_I \sqrt{1 - \left(\frac{f_N}{2\pi A_I f_I}\right)^2} - \frac{1}{\pi}\left(\frac{f_N}{f_I}\right)\cos^{-1}\left(\frac{f_N}{2\pi A_I f_I}\right) \tag{6.37}$$

A practical example of upper synchronizer and desynchronizer elastic store sizes for input sinusoidal jitter can be illustrated [6.1] for a DS1 to DS2 multiplex and using associated input jitter tolerance requirements (see Chapter 9). The worst case frequency offset due to input jitter for the low-speed DS1 input of a DS1 to DS2 multiplex is defined by the template of Figure 6.5 [6.3] and occurs at the $2A_I = 5$ UI peak-to-peak, $f_I = 500$ Hz point. Noting that the worst case synchronizer elastic store requirements are obtained for f_{P-} and f_{N-}, using the DS1 to DS2 multiplex parameter values contained in Chapter 5, and (5.12), (6.5), (6.6), (6.27), (6.30), and (6.31):

$$f_{P-} = \bar{f}_{r-} - f_{nom+} = 1545745 - 1544200 = 1545 \text{ Hz}$$

$$f_{N-} = f_{j,max} - (\bar{f}_{r+} - f_{nom,-}) = 5367 - (1545847 - 1543800) = 3320 \text{ Hz}$$

$$t_{P-,1} = 0.437 \text{ ms}$$

$$t_{N-,1} = 0.639 \text{ ms}$$

$$t_{N-,2} = 1.361 \text{ ms}$$

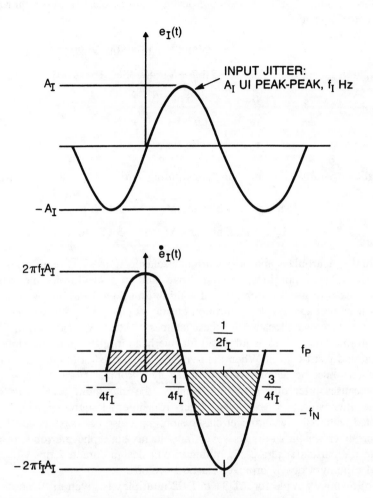

NOTE: THE MAXIMUM DESYNCHRONIZER STORE
REQUIREMENTS ARE OBTAINED FOR f_{P+} AND f_{N+}

Fig. 6.4 Effect of input jitter on desynchronizer elastic store

Similarly, noting that the worst case desynchronizer elastic store requirements are obtained for f_{P+} and f_{N+},

$$f_{P+} = \bar{f}_{r+} - f_{nom-} = 1545847 - 1543800. = 2047 \text{ Hz}$$

$$f_{N+} = f_{j,max} - (\bar{f}_{r-} - f_{nom+}) = 5367 - (1545745 - 1544200) = 3822 \text{ Hz}$$

$$t_{P+,1} = 0.416 \text{ ms}$$

$$t_{N+,1} = 0.662 \text{ ms}$$

$$t_{N+,2} = 1.338 \text{ ms}$$

Using the above parameter values, and substituting into (6.33) and (6.35)–(6.37)

$$M_{S+} = 3.55 \text{ bits and } M_{S-} = -2.13 \text{ bits}$$

$$M_{D+} = 1.88 \text{ bits and } M_{D-} = -3.22 \text{ bits}$$

The total synchronizer and desynchronizer elastic store sizes required includes both an allocation for input jitter, derived above, and an allocation for internal multiplex jitter sources (e.g., justification and waiting time, overhead bit insertion-deletion), as well as propagation delay through elastic storage cells. We will now consider the required storage allocation for these sources of jitter. As discussed in Chapter 5, the synchronizer read clock is inhibited for overhead insertion into the higher-rate output signal and justification. When the overhead is inserted at the higher-rate, f_{out}, the jitter which must be accommodated in the synchronizer store is $-f_{in}/f_{out}$ bits. Justification causes jitter of $-f_{in}/f_r$ bits. Finally, the maximum peak value of the waiting time jitter is $(f_{r+} - f_{in-})t_j$ bits.[1] Jitter from internal sources which must be accommodated within the synchronizer elastic store must also be accommodated within the desynchronizer elastic store. However, any steady-state phase error which may occur due to the particular phase-smoothing circuit design chracteristics will add to the required minimum desynchronizer elastic store size.

The following example for a DS1 to DS2 multiplex is given to illustrate these concepts [6.1].

- Synchronizer

  The maximum peak amplitude of waiting time jitter, computed in Chapter 5, is approximately $+0.38$ bits.[2] Inhibiting the read clock for DS2 overhead bits causes jitter approximately equal to -0.25 bits, and justifying the DS1 lower-rate input causes jitter approximately equal to -1.0 bits. These jitter sources may add in the synchronizer elastic store, causing jitter of $+0.38$, -1.25 bits around the justification request point. In addition, it is necessary to allow some time for propagation delay and set-up time through the elastic store before reading, and for hold time requirements after reading. Normally, an old bit may be read while a new bit is written due to propagation delay through the elastic store cells, while a new bit must have time to propagate and set-up before being

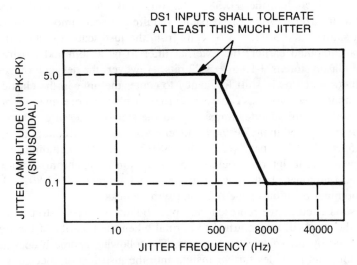

INPUT JITTER TOLERANCE TEMPLATE

DS1 INPUTS SHALL TOLERATE
AT LEAST THIS MUCH JITTER

NOTE: THIS REQUIREMENT SHALL BE EFFECTIVE
AS OF APRIL 1, 1984

Fig. 6.5 DS1 to DS2 multiplex input jitter tolerance requirement

read. In this example, delay uncertainties can be allowed for by allocating −0.25 bits in the store for timing margin.

- Desynchronizer
The desynchronizer elastic store must accommodate −1.0 bit due to removal of justification bits, −0.25 bit due to removal of the higher order overhead bits from the frame format, and an allowance of −0.25 bit to provide timing margin. In the desynchronizer elastic store, the nominal position of the read and write clocks is determined by the requirements of the phase-smoothing circuit. Typically, a phase-locked loop will be designed for a steady-state phase error equal to zero when the average positions of the read and write clocks are separated by half the total elastic store capacity. Any steady-state phase error which may occur due to the characteristics of the phase-locked loop design will add to the required minimum capacity of the desynchronizer elastic store. These phase errors tend to be symmetrical; in this example let ±0.5 bit additional capacity be allowed for steady-state phase error [6.1]. Thus, a total minimum capacity of ±1.25 bits around the average position of the read and write clocks is necessary in the desynchronizer elastic store to accommodate only the internal sources of jitter.

We may now compute the *total* synchronizer and desynchronizer elastic store sizes required for external and internal jitter sources. As computed earlier in this chapter, the synchronizer elastic store would experience a movement of the read storage location relative to the write storage location of $+3.55$, -2.13 bits peak due to external jitter. Including the $+0.38$, -1.5 bits peak due to internal sources, the minimum synchronizer elastic store capacity required to accommodate both internal and external sources is $+3.93$, -3.63 cells from the justification request threshold. Similarly, as computed earlier, the external jitter to be accommodated by the desynchronizer elastic store is $+1.88$, -3.22 bits. However, the desynchronizer phase-locked oscillator will slowly shift frequency to center the jitter in the elastic store if the applied input jitter persists. The desynchronizer elastic store must therefore accommodate both the initial jitter transient and the steady-state case. Then, the additional capacity required in the desynchronizer elastic store to accommodate external jitter is $+2.55$, -3.22 bits peak $[(3.22 + 1.88)/2 = 2.55)]$. Including the $+1.25$, -1.25 bits peak due to internal sources, the minimum desynchronizer elastic store capacity to accommodate both internal and external sources is $+3.80$, -4.47 bits around the nominal position of the read and write clocks.

The results cited above represent an upper bound on elastic store size as they assume that the store fill has returned to normal when the polarity of the frequency deviations caused by sinusoidal jitter reverses. The following approach does not make this assumption, and provides further insight into the justification mechanism [6.4]. It examines phase accumulation in the synchronizer elastic store due to sinusoidal input jitter utilizing an iterative approach, which leads to a steady state solution.

Consider again the scenario in which $2\pi A_I f_I$ is greater than both f_P and f_N. Referring to Figure 6.6a, the positive excursion of the synchronizer elastic store during the time interval from $t = 0$ to $t = t_{P,1}$ is given by $M_{S+}/2$, where M_{S+} is defined in (6.33), and will be denoted $M_{S,a}$. Note that $f_{net}(t)$ turns negative upon crossing time $t_{P,1}$, where the positive deviation from the nominal becomes less than f_P. Figure 6.6b illustrates the excursion of the elastic store from its *normal* fill point, $M_S = 0$. We see the elastic store remains above its normal fill until time $t_{0,1}$ though the positive excursion is decreasing. Therefore, during this interval, for the justification algorithm under discussion, even though $\dot{e}_I(t) < f_P$, the multiplex will not resume justification.

The accumulated negative phase (in terms of bits) during this interval ($t_{P,1} \le t < t_{0,1}$) is

$$M_{S,b} = \int_{t_{P,1}}^{t_{0,1}} \{\dot{e}_I(t) - f_P\}dt$$

$$M_{S,b} = A_I\left\{ \sin 2\pi f_I t_{0,1} - \sqrt{1 - \left(\frac{f_P}{2\pi A_I f_I}\right)^2} \right\}$$

$$+ \frac{1}{2\pi}\left(\frac{f_P}{f_I}\right) \cos^{-1}\left(\frac{f_P}{2\pi A_I f_I}\right) - f_P t_{0,1} \tag{6.38}$$

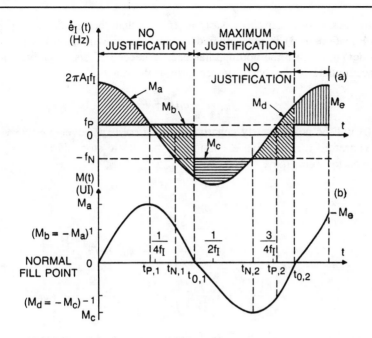

(a) THE SHADED AREAS ILLUSTRATE THE ACCUMULATED
PHASE IN THE SYNCHRONIZER STORE FOR ONE CYCLE
OF THE INPUT JITTER

(b) AN EXAMPLE OF THE IMPACT OF THIS ACCUMULATED
PHASE ON THE SYNCHRONIZER STORE FILL IS
EXPLICITLY ILLUSTRATED

NOTE: THESE GRAPHS ARE BASED ON INPUT SINUSOIDAL
JITTER OF PEAK AMPLITUDE $A_I = 2.5$ UI, AND
FREQUENCY $f_I = 500$ Hz

Fig. 6.6 Impact of sinusoidal input jitter on synchronizer elastic store fill

where the time $t_{0,1}$ is defined by

$$M_{S,a} + M_{S,b} = 0 \qquad (6.39)$$

or

$$A_I \sin(2\pi f_I t_{0,1}) - f_P t_{0,1} = 0 \qquad (6.40)$$

For the scenario illustrated in Figure 6.6, $t_{0,1} > t_{N,1}$. In this case, when the elastic store returns to its normal fill point at $t_{0,1}$, the magnitude of $\dot{e}_I(t) > f_N$. This indicates that even when justification is at the maximum rate, the rate at which data is being written into the synchronizer elastic store is still less than the rate at which it is being read out. Equation (6.3) is not satisfied because the maximum tolerable negative frequency deviation from the nominal has been exceeded. Thus, negative phase will begin to accumulate in the synchronizer store at the rate

$$f_{net}(t) = \dot{e}_I(t) - (-f_N)$$

Thus, $f_{net}(t)$ will be continuously negative while $|\dot{e}_I(t)| > f_N$, over the interval $t_{0,1} \leq t < t_{N,2}$. Therefore the negative excursion of the elastic store during this interval $M_{S,c}$, is given by (see Figure 6.6)

$$M_{S,c} = \int_{t_{0,1}}^{t_{N,2}} \{\dot{e}_I(t) + f_N\}dt$$

$$= A_I\left\{ -\sqrt{1 - \left(\frac{f_N}{2\pi A_I f_I}\right)^2} - \sin(2\pi f_I t_{0,1}) \right\}$$

$$+ \left\{ \frac{1}{2}\left(\frac{f_N}{f_I}\right) + \frac{1}{2\pi}\left(\frac{f_N}{f_I}\right) \cos^{-1}\left(\frac{f_N}{2\pi A_I f_I}\right) - f_N t_{0,1} \right\} \qquad (6.41)$$

The situation after time $t_{N,2}$ is similar to that after $t_{P,1}$. We see the synchronizer elastic store remains below its normal fill until time $t_{0,2}$, though the magnitude of the negative excursion is decreasing. Therefore, during this interval, for the justification algorithm under discussion, even though $|\dot{e}_I(t)| < f_N$, the multiplex will not cease justification.

The accumulated positive phase during this interval ($t_{N,2} \leq t < t_{0,2}$) is

$$M_{S,d} = \int_{t_{N,2}}^{t_{0,2}} \{\dot{e}_I(t) + f_N\}dt$$

$$= A_I\left\{ \sin(2\pi f_I t_{0,2}) + \sqrt{1 - \left(\frac{f_N}{2\pi A_I f_I}\right)^2} \right\}$$

$$- \frac{1}{2}\left(\frac{f_N}{f_I}\right) - \frac{1}{2\pi}\left(\frac{f_N}{f_I}\right) \cos^{-1}\left(\frac{f_N}{2\pi A_I f_I}\right) + f_N t_{0,2} \qquad (6.42)$$

where the time $t_{0,2}$ is defined by

$$M_{S,c} + M_{S,d} = 0 \qquad (6.43)$$

or

$$A_I\{\sin(2\pi f_I t_{0,2}) - \sin(2\pi f_I t_{0,1})\} + f_N(t_{0,2} - t_{0,1}) = 0 \qquad (6.44)$$

For the scenario illustrated in Figure 6.6, $t_{0,2} > t_{P,2}$. As $\dot{e}_I(t_{0,2}) > f_P$, (6.3) is not satisfied because the maximum tolerable positive frequency deviation from the nominal has been exceeded. Thus, positive phase will begin to accumulate in the synchronizer elastic store at the rate

$$f_{net}(t) = \dot{e}_I(t) - f_P$$

$f_{net}(t)$ will be continuously positive while $\dot{e}_I(t) > f_P$ over the interval $t_{0,2} \leq t < 1/f_I$, therefore the positive excursion of the elastic store over this interval, $M_{S,e}$, is given by (see Figure 6.6)

$$M_{S,e} = \int_{t_{0,2}}^{1/f_I} \{\dot{e}_I(t) - f_P\}dt$$

$$M_{S,e} = A_I\{-\sin(2\pi f_I t_{0,2})\} - f_P\left(\frac{1}{f_I} - t_{0,2}\right) \tag{6.45}$$

Clearly, this is an iterative process, for which a steady-state solution will ultimately be obtained. Note that for the special case examined, in which $2\pi A_I f_I$ is larger in magnitude than both f_P and f_N, the multiplexer alternated between maximum justification and no justification. Also note that in the limiting case of very large frequency deviations compared to the maximum justification rate, $2\pi A_I f_I/f_P \to \infty$ and $2\pi A_I f_I/f_N \to \infty$, all the jitter must be accommodated in the synchronizer elastic store. Again, the applied sinusoidal input jitter which is encoded by the justification mechanism is reproduced exactly at the input to the desynchronizer elastic store.

The remainder of this section briefly outlines considerations regarding the jitter tolerance of slip buffering and pointer adjustment synchronization multiplexes.

- As discussed in Chapter 5, a slip buffering multiplex obtains its timing from a source independent of the incoming signal (but synchronous or plesiochronous with it). To prevent slips from occurring in the synchronizer elastic store, the number of elastic storage cells, M_S, must clearly be at least as large as the magnitude of the relative peak jitter between the extracted and independent clocks, $|e_{rel}(t)|_{peak}$. In addition, some extra allowance, Δ, must normally be included to account for elastic storage cell delay margin. Assuming that the read and write clocks are initially phased $M_S/2$ cells apart when each is jitter-free we then require that [6.2]

$$\frac{M_S}{2} - \Delta \geq |e_{rel}(t)|_{peak} \tag{6.46}$$

- As discussed in Chapter 5, the pointer adjustment mechanism is analogous to a positive-negative justification mechanism. The principles developed in this section may be used to analyze negative justification as well. It is then possible to extend these analyses to obtain relationships for positive-negative and positive-zero-negative justification. We have not addressed this extension within this section.

6.1.2 Multiplex Jitter Tolerance Measurements

As discussed earlier, the jitter tolerance of digital multiplexes is dependent upon the input timing recovery circuit tolerance, synchronizer and desynchronizer elastic store sizes, and/or justification or pointer adjustment capacity. The transmission penalty criterion, described in Chapter 4, allows environment independent determination of the decision circuit alignment jitter allocation, which is critical for evaluating input clock extraction circuit tolerance to input jitter. This section, based on [6.5], describes the onset of errors criterion for jitter tolerance, which is recommended for evaluating those factors unrelated to input clock extraction circuit tolerance.

The onset of errors criterion for jitter tolerance measurements is defined as the largest amplitude of jitter at a specified frequency that causes a cumulative total of no more than 2 errored seconds, where these errored seconds have been summed over successive 30 second measurement intervals of increasing jitter amplitude.

This technique involves setting a jitter frequency and determining the jitter amplitude of the test signal which causes the onset of errors criterion to be satisfied. Specifically, this technique requires: (1) isolation of the jitter amplitude "transition region" (in which error-free operation ceases), (2) one errored second measurement, 30 seconds in duration, for each incrementally increased jitter amplitude from the beginning of this region, and (3) determination of the largest jitter amplitude for which the cumulative errored second count is no more than 2 errored seconds. The examples in Table 6.1 (using actual measurement data) illustrate how to determine the jitter amplitude which satisfies the above criterion. The process is repeated for a sufficient number of frequencies such that the measurement accurately represents the continuous sinusoidal input jitter tolerance of the multiplex over the applicable jitter frequency range. The test equipment must be able to produce a controlled jittered signal and measure the resulting errored seconds caused by the jitter on the incoming signal.

Test Configuration

Figure 6.7 illustrates the test configuration for the onset of errors technique. The optional frequency synthesizer is used to provide a more accurate determination of frequencies utilized in the measurement procedure. The optional jitter receiver is used to verify the amplitude of generated jitter.

Procedure

1. Connect the equipment as shown in Figure 6.7. Verify proper continuity and error-free operation.

2. Set the input jitter frequency as desired, and initialize the jitter amplitude to zero unit intervals peak-to-peak.

3. Increase the jitter amplitude in gross increments to determine the amplitude region where error-free operation ceases. Reduce the jitter amplitude to its level at the beginning of this region.

4. Record the number of errored seconds that occur over a 30 second measurement interval. Note that the initial measurement must be zero errored seconds.

5. Increase the jitter amplitude in fine increments, repeating Step 4 for each increment, until the onset of errors criterion is satisfied.

6. Record the indicated amplitude and frequency of the applied input jitter, and repeat Steps 2–6 for a sufficient number of frequencies to characterize the jitter tolerance curve.

Table 6.1 Determination of Jitter Amplitude Satisfying Defined Onset of Errors Criterion

Sinusoidal Jitter Amplitude UI Peak-Peak	Example 1		Example 2		Example 3		Example 4	
	Determined Amplitude	ES*	Determined Amplitude	ES	Determined Amplitude	ES	Determined Amplitude	ES
07.50		0		0		0		0
07.55		0		0		0		0
07.60		0		0		1		0
07.65		0		0		0		0
07.70		0		0		0		0
07.75		0		0		0		0
07.80		0		1		0		1
07.85		0	7.85	1		0		0
07.90		0		1		0		0
07.95		1		1	7.95	0	7.95	1
08.00	8.00	0		1		3		1
08.05		4		2		6		1

*Errored seconds.

6.2 MULTIPLEX JITTER TRANSFER AND MEASUREMENTS

Multiplex jitter transfer characteristics are specified within 2.048 Mb/s based hierarchies, whereas demultiplexer jitter transfer characteristics are specified within 1.544 Mb/s based hierarchies (see Chapter 9). The intent of both specifications is to place requirements on the desynchronizer jitter transfer characteristic. Desynchronizer jitter transfer requirements commonly restrict the amount of peaking and the corner frequency of the desynchronizer phase-smoothing circuit, as illustrated in Figure 6.8,

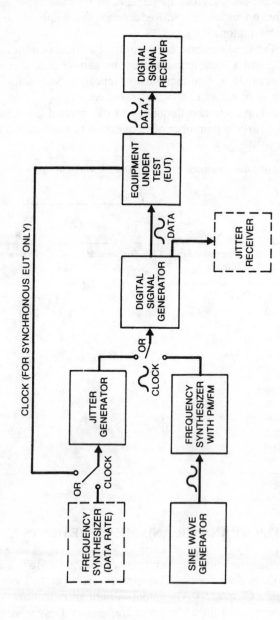

Fig. 6.7 Jitter tolerance measurement configuration: onset of errors technique

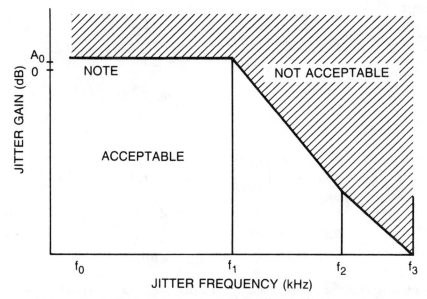

Fig. 6.8 Desynchronizer jitter transfer requirement template

to ensure proper attenuation of both inherent and input multiplex jitter. Inadequate attenuation may result in excessive accumulation within a digital network, which will eventually degrade its performance. Multiplex jitter accumulation is discussed further in Chapter 7.

Compliance with desynchronizer jitter transfer characteristic specifications is generally determined by applying sinusoidal jitter of a selected amplitude and frequency to the multiplex lower-rate signal input or the demultiplexer higher-rate signal input and observing the jitter amplitude at the demultiplexer lower-rate signal output at the applied frequency. In order to accurately determine the desynchronizer jitter transfer characteristic, the impact of sinusoidal jitter applied to the standard multiplex input ports must be examined further.

6.2.1 Impact of Input Jitter Parameters on Jitter Transfer

A general expression for multiplex output jitter generation in the presence of input jitter is discussed in Chapter 7, and is reproduced below.

$$\Phi_S(f) = \text{sinc}^2(f)\tilde{Q}(f) + \sum_{n=1}^{\infty} \left(\frac{\rho}{2\pi n}\right)^2$$

$$\cdot \{\delta(f - nf_{j,max}) + \delta(f + nf_{j,max})\}$$

$$+ \text{sinc}^2(f) \sum_{k=-\infty}^{\infty} \Phi_I(f - kf_{j,max}) \tag{6.47}$$

where

$$\tilde{Q}(f) = \sum_{n=1}^{\infty} \left(\frac{1}{2\pi n}\right)^2 \left\{ \sum_{k=-\infty}^{\infty} Z_n(f - n\rho f_{j,max} - kf_{j,max}) \right.$$

$$+ Z_n(-f - n\rho f_{j,max} - kf_{j,max}) \tag{6.48}$$

$$Z_n(f) = F\{Z_n(t)\} \tag{6.49}$$

$$z_n(t) = E[\exp(-j2\pi n (e_I(t) - e_I(0)))] \tag{6.50}$$

This expression illustrates that the jitter at a demultiplexer low-speed output, when jitter is applied to the corresponding multiplexer low-speed input, is effectively made up of two terms:

- A feed-through term dependent on the filtered replicated input jitter spectrum.
- A discrete spectra of "distortion" products.

We may derive an expression for multiplex output jitter in the presence of sinusoidal input jitter, of peak amplitude A_I and frequency f_I using the above expression. As discussed in [6.6], sinusoidal jitter may be modeled by the random process

$$e_I(t) = A_I \sin(2\pi f_I t + \theta) \tag{6.51}$$

where θ is a random variable distributed uniformly on $[0, 2\pi)$. We may then obtain expressions for $z_n(t)$, $Z_n(f)$, and $\Phi_I(f)$. Substituting into (6.50),

$$z_n(t) = E[e^{-j2\pi nA_I \sin(2\pi f_I t + \theta)} e^{j2\pi nA_I \sin\theta}]$$

Using the expression

$$e^{j\lambda \sin\chi} = \sum_{k=-\infty}^{\infty} J_k(\lambda)e^{jk\chi}$$

where $J_k(\lambda)$ represent Bessel functions of the first kind, we have

$$z_n(t) = E\left[\sum_{k=-\infty}^{\infty} J_k(2\pi nA_I)e^{-jk(2\pi f_I t + \theta)} \sum_{l=-\infty}^{\infty} J_l(2\pi nA_I)e^{jl\theta}\right]$$

$$= \sum_{k=-\infty}^{\infty} \sum_{l=-\infty}^{\infty} J_k(2\pi nA_I)J_l(2\pi nA_I)e^{-jk2\pi f_I t}E[e^{-j(k-l)\theta}]$$

$$= \sum_{k=-\infty}^{\infty} J_k^2(2\pi nA_I)e^{-jk2\pi f_I t} \tag{6.52}$$

Using (6.49)

$$Z_n(f) = F\{z_n(t)\} = \int_{-\infty}^{\infty} \sum_{k=-\infty}^{\infty} J_k^2(2\pi nA_I)e^{-jk2\pi f_I t}e^{-j2\pi ft}dt$$

$$= \sum_{k=-\infty}^{\infty} J_k^2(2\pi nA_I)\delta(f + kf_I) \tag{6.53}$$

$\Phi_I(f)$ may be determined by taking the Fourier transform of the covariance $C_I(t)$, where

$$C_I(t) = E[\{e_I(t + s) - \mu_I\}\{e_I(s) - \mu_I\}]$$

where $\mu_I = E\{e_I(t)\}$ and $E[e_I(\cdot)]$ is independent of $(\cdot)$. Furthermore, as we have a stationary process, we may choose $s = 0$. Then

$$E[e_I(\cdot)] = E[e_I(0)] = E[A_I \sin \theta] = 0$$

and $C_I(t)$ may be simplified

$$C_I(t) = E[A_I \sin(2\pi f_I t + \theta)A_I \sin \theta]$$

$$= A_I^2 \frac{1}{2\pi} \int_0^{2\pi} \sin(2\pi f_I t + \theta)\sin \theta \, d\theta$$

$$= \frac{A_I^2}{2\pi} \sin(2\pi f_I t) \int_0^{2\pi} \sin \theta \cos \theta \, d\theta$$

$$+ \frac{A_I^2}{2\pi} \cos(2\pi f_I t) \int_0^{2\pi} \sin^2 \theta \, d\theta$$

$$= \frac{A_I^2}{2} \cos(2\pi f_I t) \tag{6.54}$$

Therefore,

$$\Phi_I(f) = F\{C_I(t)\} = \int_{-\infty}^{\infty} \frac{A_I^2}{2} \cos(2\pi f_I t) e^{-j2\pi f t} dt$$

$$= \frac{A_I^2}{4} \{\delta(f - f_I) + \delta(f + f_I)\} \tag{6.55}$$

Combining terms

$$\Phi_s(f) = \text{sinc}^2(f) \sum_{n=1}^{\infty} \frac{1}{(2\pi n)^2} \sum_{m=-\infty}^{\infty} \sum_{k=-\infty}^{\infty} J_k^2(2\pi n A_I)$$

$$\cdot \{\delta(f - n\rho f_{j,max} + k f_I - m f_{j,max})$$

$$+ \delta(f + n\rho f_{j,max} + k f_I - m f_{j,max})\}$$

$$+ \sum_{n=1}^{\infty} \left(\frac{\rho}{2\pi n}\right)^2 \{\delta(f - n f_{j,max}) + \delta(f + n f_{j,max})\}$$

$$+ \text{sinc}^2(f) \frac{A_I^2}{4} \sum_{k=-\infty}^{\infty} \{\delta(f - f_I - k f_{j,max}) + \delta(f + f_I - k f_{j,max})\} \tag{6.56}$$

Assuming that the passband of the desynchronizer jitter transfer characteristic, $G(f)$, for a typical multiplex and the input sinusoidal jitter frequency, f_I, are small compared to the maximum justification rate, (6.56) may be simplified by neglecting the second term of $\Phi_s(f)$, as well as all terms in the replicated feed-through term except the $k = 0$ term. Then $\Phi_s(f)$ may be expressed as

$$\Phi_s(f) = \text{sinc}^2(f)\Phi_d(f) + \text{sinc}^2(f)\Phi_{f-t}(f) \tag{6.57}$$

where

$$\Phi_d(f) = \sum_{n=1}^{\infty} \frac{1}{(2\pi n)^2} \sum_{m=-\infty}^{\infty} \sum_{k=-\infty}^{\infty} J_k^2(2\pi n A_I)\{\delta(f - n\rho f_{j,max} + k f_I - m f_{j,max})$$

$$+ \delta(f + n\rho f_{j,max} + k f_I - m f_{j,max})\} \tag{6.58}$$

and

$$\Phi_{f-t}(f) = \frac{A_I^2}{4} \sum_{k=-\infty}^{\infty} \{\delta(f - f_I) + \delta(f + f_I)\} \tag{6.59}$$

Note the effect of slightly varying the input frequency on the measured jitter transfer

haracteristic in the example illustrated in Figure 6.9 [6.7]; this example demon-
strates the sensitivity of the multiplex transfer characteristic measurement to the dis-
ortion products spectra, which are modulated by this variation.

The ratio of the filtered distortion power, σ_d^2, to the filtered feed-through power,
σ_{f-t}^2, can be expressed as

$$\frac{\sigma_d^2}{\sigma_{f-t}^2} = \frac{\int_{-\infty}^{\infty} |G(f)|^2 \, \mathrm{sinc}^2(f)\Phi_d(f)df}{\int_{-\infty}^{\infty} |G(f)|^2 \, \mathrm{sinc}^2(f)\frac{A_I^2}{4}\{\delta(f-f_I) + \delta(f+f_I)\}df} \tag{6.60}$$

NOTES:
(1) $A_I = 1.0$ UI PEAK-PEAK: $f_{nom} = 44.7360$ Mb/s
(2) $A_I = 1.0$ UI PEAK-PEAK: $f_{nom} = 44.7351$ Mb/s
(3) $A_I = 1.0$ UI PEAK-PEAK: $f_{nom} = 44.7369$ Mb/s

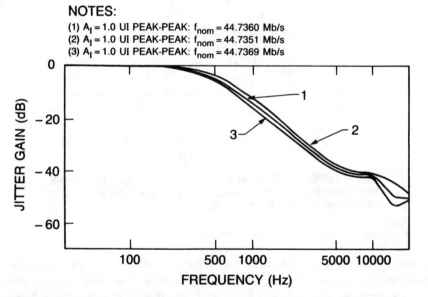

Fig. 6.9 Multiplex jitter transfer characteristic

This ratio is plotted in Figure 6.10 [6.7, 6.8] as a function of sinusoidal input amplitude and frequency for a DS1 to DS2 multiplex having nominal justification ratio
and desynchronizer phase-smoothing circuit characteristics. For a given input sinusoidal jitter frequency, the relative distortion power decreases with input amplitude
but increases with frequency. (This former effect is analogous to quantization distortion for a uniform quantizer. The high distortion ratio for small input amplitudes
is analogous to the situation which causes idle channel noise [6.8].) The frequency
deviation limit for a DS1 to DS2 multiplex is also plotted; as can be seen from Figure
6.10, measurement accuracy of a few tenths of a dB cannot be obtained. However,
(6.58) shows that unless the justification ratio is an exact integer ratio (i.e., $\rho = i/$

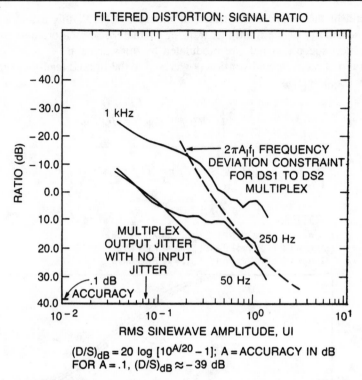

FILTERED DISTORTION: SIGNAL RATIO

$$(D/S)_{dB} = 20 \log [10^{A/20} - 1]; \quad A = \text{ACCURACY IN dB}$$
$$\text{FOR } A = .1, \ (D/S)_{dB} \approx -39 \text{ dB}$$

NOTE: THESE CURVES ARE FOR A DS1 TO DS2 MULTIPLEX
WITH NOMINAL JUSTIFICATION RATIO AND TYPICAL PHASE
SMOOTHING CIRCUIT PARAMETERS

Fig. 6.10 Filtered distortion to signal ratio constraints

j, where i,j are integers) the distortion terms will not "fold over" on top of the original input frequency. Even if the justification ratio is an exact integer ratio, but i and j are large, then the resultant "fold over" terms will be negligible both because of the $1/n^2$ weighting and the factor of n in the Bessel function of (6.58). The result implies that as long as a very narrowband measurement is used, the noncoherent distortion productions can in general be filtered out, and the filtered feed-through power can be measured [6.8].

Another consideration for jitter transfer is the frequency deviation caused by sinusoidal input jitter. As described in Section 6.1, sinusoidal input jitter producing frequency deviations in excess of f_P and/or f_N will result in saturation of the justification mechanism, so that not all of the input jitter is transferred to the demultiplexer. The impact on measured jitter transfer from multiplexer lower-rate input to demultiplexer lower-rate output in this cases is illustrated in Figure 6.11 [6.7, 6.8].

If characterization of the desynchronizer jitter transfer function is desired utilizing the multiplexer lower-rate input and demultiplexer lower-rate output ports, it

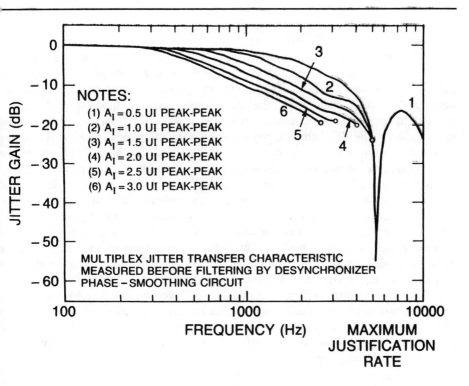

Fig. 6.11 Impact of input jitter amplitude on measured multiplex jitter transfer

s necessary to "linearize" the multiplexing process by applying appropriate constraints to the applied input jitter amplitude and frequency. Considering the analyses contained in Section 6.1, the amplitude for a given frequency should be small enough o avoid saturating the justification mechanism, but sufficiently large to minimize he relative distortion power. As illustrated in Figure 6.10 [6.7, 6.8], these constraints are conflicting at higher frequencies.

When sinusoidal jitter modulates the phase of the input signal to the demultiplexer, the output jitter spectrum contains a discrete component at the frequency of the input jitter, in addition to the inherent multiplex jitter components already present. If characterization of desynchronizer jitter transfer is desired using the demultiplexer higher-rate and lower-rate ports, the amplitude of the applied input jitter should be sufficiently large to ensure that its contribution to the output jitter spectrum at the applied frequency dominates that of the inherent multiplex jitter, but small enough that it does not exceed the demultiplexer input jitter tolerance.

6.2.2 Multiplex Jitter Transfer Measurements

We will describe two techniques, based on [6.5], which allow the determination of the desynchronizer phase-smoothing circuit transfer function using standard mult plex interfaces.

The first technique determines the desynchronizer jitter transfer characterist using the multiplexer lower-rate and demultiplexer lower-rate input ports. Sinusoid: jitter of a selected amplitude and frequency is applied to the multiplexer low-spee input, and the jitter amplitude at the demultiplexer low-speed output is observed : the applied frequency. The process is repeated for a sufficient number of frequencie to characterize the desynchronizer jitter transfer function. Specifically, when sinu soidal jitter modulates the phase of the input signal to one of the multiplexer low speed inputs, the jitter spectrum appearing at the corresponding tributary output, i addition to containing other inherent multiplex jitter components at discrete location throughout the spectrum, contains a discrete component at the frequency of the inpu jitter. This technique involves making the amplitude of the input jitter sufficient] large to ensure that this discrete component in the output jitter spectrum at the applie frequency dominates the other inherent multiplex jitter components in the measure ment bandwidth. However, it should not be so large as to saturate the multiplexe justification mechanism (onset of saturation).

To achieve a high degree of accuracy, the spectrum analyzer bandwidth mu be sufficiently narrow to obtain the desired amplitude resolution and dynamic rang in each frequency band measured. It is also assumed that the transfer function of th multiplexer low-speed input clock extraction circuit does not alter the applied jitte in the frequency range of interest.

Test Configuration: Multiplex Technique

Figure 6.12 [6.5] illustrates the test configuration for the jitter transfer functio measurement. The optional frequency synthesizer may be used to provide a mor accurate determination of frequencies utilized in the measurement procedure.

Procedure

1. Perform a jitter tolerance measurement over the desired frequency range.
2. Connect the equipment as shown in Figure 6.12, bypassing the multiplex. Ver ify proper continuity, linearity, and error-free operation.
3. Manually set the test frequency on the spectrum analyzer.
4. Adjust the tracking oscillator output level on the spectrum analyzer to produc the largest tolerable jitter amplitude which will not cause onset of saturatio (as defined in this section) at the selected frequency.
5. Set the spectrum analyzer bandwidth as narrow as is feasible, and record th zero dB amplitude transfer reference level of the test equipment.
6. Reconnect the multiplex as shown in Figure 6.12. Verify proper continuity an error-free operation.

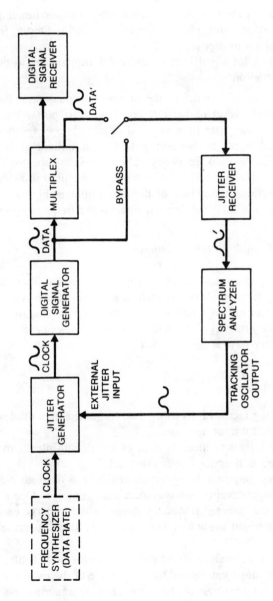

Fig. 6.12 Jitter transfer function measurement configuration: multiplex technique

7. Record the magnitude of the overall (test equipment and multiplex) jitter transfer function. Averaging is generally required to remove the effects of inherent multiplex jitter on the measurement.
8. To obtain the magnitude of the multiplex jitter transfer function, subtract the zero dB amplitude transfer reference level, recorded in Step 5, from the overall magnitude obtained in Step 7.
9. Repeat Steps 3–8 for a sufficient number of frequencies to characterize the jitter transfer function.

The second technique determines the desynchronizer jitter transfer characteristic utilizing the demultiplexer higher-rate and lower-rate ports. This technique involves applying sinusoidal jitter of a selected amplitude and frequency to the demultiplexer high-speed input, and observing the jitter amplitude at the demultiplexer low-speed output at the applied frequency. The process is repeated for a sufficient number of frequencies to characterize the desynchronizer jitter transfer function. We also assume that the transfer function of the demultiplexer high-speed input clock extraction circuit does not alter the applied jitter in the frequency range of interest.

Test Configuration: Demultiplexer Technique

Figure 6.13 [6.5] illustrates the test configuration for the jitter transfer function measurement utilizing the demultiplexer technique. It should be emphasized that the following procedure *can not calibrate out the effects of the low-speed receiver circuitry contained in the jitter receiver functional block component,* and therefore requires that this circuitry have a flat response.

Procedure

1. Calculate a scaling factor in dB using the ratio of the demultiplexer high-speed input to low-speed output data rates.
2. Perform a jitter tolerance measurement of the demultiplexer over the desired frequency range, as described in Section 6.1.2.
3. Set the frequency range on the spectrum analyzer as desired. Adjust the tracking oscillator output level on the spectrum analyzer to produce a tolerable jitter amplitude over the selected frequency range, which is large enough to ensure adequate measurement accuracy, yet sufficiently small to preserve linear operation.
4. Setting the spectrum analyzer bandwidth as narrow as feasible, sweep the selected frequency range and record the magnitude of the overall (test equipment and demultiplexer) jitter transfer function. (Setting a narrow spectrum analyzer bandwidth may allow a reduction in applied jitter amplitude with no loss in measurement accuracy.)

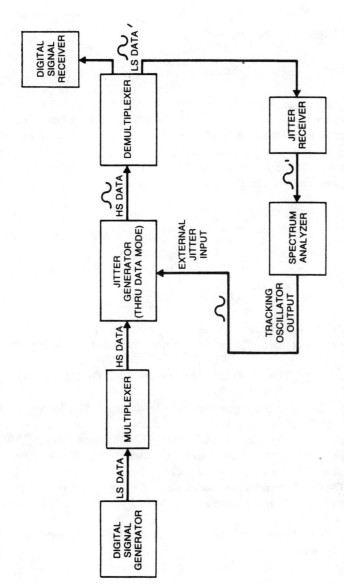

Fig. 6.13 Jitter transfer function measurement configuration: demultiplexer technique

5. To obtain the demultiplexer jitter transfer function, subtract the scaling factor in dB determined in Step 1 from the overall jitter transfer function recorded in Step 4.

6. Repeat Steps 2–5 for a sufficient number of frequency ranges to characterize the overall frequency range of interest.

6.3 REMARKS

Within this chapter we have examined the jitter tolerance and transfer of digital multiplexes. Relationships between input jitter tolerance and multiplex parameters have been derived, and a technique for measuring multiplex jitter tolerance has been described. Multiplex jitter transfer characteristics have been examined, taking into account multiplex tolerance implications, and two techniques for determining the desynchronizer jitter transfer characteristic have been described.

NOTES

1. Here we have neglected t_{thr}, discussed in Chapter 5.
2. For simplicity, we have ignored input jitter in this computation.

REFERENCES

[6.1] C. E. Huffman, J. K. Blake, "Asynchronous Multiplex Jitter," *Collins Transmission Systems Division Technical Bulletin,* 523-0605721-00283J, pp. 6-1–6-18.

[6.2] R. J. S. Bates, "A Model for Jitter Accumulation in Digital Networks," *Globecom 1983,* pp. 145–149, 1983.

[6.3] "Digital Multiplexes, Requirements and Objectives," *Bell System Technical Reference,* PUB 43802, July 1982.

[6.4] D. L. Husa, G. R. Ritchie, E. L. Varma, "Unpublished work on Multiplex Elastic Store Behavior for Sinusoidal Input Jitter Using a Phase Model and its Comparison to a Frequency Model," AT&T Bell Laboratories, 1982.

[6.5] T1X1.4/87-024, "Jitter Measurement Methodology," *T1 Committee,* 1987.

[6.6] D. L. Duttweiler, "Waiting Time Jitter," *BSTJ* Vol. 51, No. 1, January 1972, pp. 165–207.

[6.7] T1X1.3/86-088, "Characterization Guidelines: Desynchronizer Jitter Transfer," *T1 Committee,* 1986.

[6.8] R. J. S. Bates, "Unpublished Work on Jitter Transfer Characterization," AT&T Bell Laboratories, 1982.

[6.9] E. L. Varma, T1X1.3 Jitter and Wander Subworking Group contribution JWC-85-100, "Demultiplexer Jitter Transfer Characteristic Specification," *T1 Committee,* 1985.

Chapter 7
Jitter Accumulation in Digital Networks

Understanding jitter accumulation in digital networks is growing in importance for several reasons. There is an increasing digital connectivity of the network, an evolution of the network toward synchronous operation, and an expanding offering of digital services. We discussed jitter accumulation from cascaded line regenerators in Chapter 3. In this chapter, jitter accumulation from cascaded bit justification multiplexes is analyzed, and a model is described that can be used to compute jitter accumulation in digital networks comprised of cascaded bit justification multiplexes and line regenerators. In addition, we will discuss the implications of coupling synchronous optical networks utilizing pointer processing network elements with the existing digital network.

7.1 JITTER ACCUMULATION FROM CASCADED BIT JUSTIFICATION MULTIPLEXES

Analysis of jitter accumulation from cascaded bit justification multiplexes is made more complex by the nonlinear interaction of input jitter with the multiplex justification process. In this section we examine jitter accumulation and its statistical distribution for cascaded bit justification multiplexes.

The analysis of multiplex jitter using random processes, described in Chapter 5, may be generalized to find the multiplex output jitter in the presence of input jitter [7.1]. It is again convenient in the following derivation to express time in terms of justification opportunities and frequency in cycles per justification opportunity.

We denote $e_I(t)$ as the jitter on the signal at the input to the synchronizer, $e_S(t)$ as the jitter on the output of the synchronizer, and $e_D(t)$ as the jitter on the output of the desynchronizer. As the synchronizer phase comparator output, and the jitter on the gapped synchronizer read clock output are no longer identical in the presence of input jitter, it is necessary to differentiate between them. Identifying the output of the synchronizer phase comparator by $e_{SPC}(t)$,

$$e_S(t) = e_{SPC}(t) + e_I(t) \tag{7.1}$$

Because of input jitter, the instantaneous frequency of the input to the synchronizer is

$$f(t) = f_{nom} + \dot{e}_I(t) \tag{7.2}$$

Thus, the instantaneous justification ratio, $\rho(t)$, is

$$\rho(t) = \bar{f}_r - f(t) \tag{7.3}$$

$$= \bar{f}_r - f_{nom} - \dot{e}_I(t) = \rho - \dot{e}_I(t) \tag{7.4}$$

We introduce the random variables ξ and τ, as defined in Chapter 5. If no justification occurs, $e_{SPC}(t)$ may be represented as

$$e_{SPC}(t) = (\Lambda - 1) + \xi + \int_{-\tau}^{t} (\rho - \dot{e}_I(s))ds \tag{7.5}$$

When justification is accounted for, the output of the synchronizer phase comparator can be expressed as

$$e_{SPC}(t) = (\Lambda - 1) + \xi + \int_{-\tau}^{t} (\rho - \dot{e}_I(s))ds$$
$$- \left[\xi + \int_{-\tau}^{[t+\tau]-\tau} (\rho - \dot{e}_I(s))ds \right] \tag{7.6}$$

Using (7.1)

$$e_S(t) = (\Lambda - 1) + \xi + \int_{-\tau}^{t} (\rho - \dot{e}_I(s))ds$$
$$- \left[\xi + \int_{-\tau}^{[t+\tau]-\tau} (\rho - \dot{e}_I(s))ds \right] + e_I(t) \tag{7.7}$$

The covariance of $e_S(t)$ has been calculated in [7.1], and is given by

$$C_S(t) = C_{S1}(t) + C_{S2}(t) + C_{S3}(t) \tag{7.8}$$

where

$$C_{S1}(t) = A(t) * \left(r(t) \sum_{k=-\infty}^{\infty} \delta(t - k) \right) \tag{7.9}$$

$$C_{S2}(t) = E[\rho^2 v(t + \tau) v(\tau)] = \rho^2 w(t) \tag{7.10}$$

$$C_{S3}(t) = A(t) * \left(C_I(t) \sum_{k=-\infty}^{\infty} \delta(t - k) \right) \tag{7.11}$$

in which $r(t)$ is defined as

$$r(t) = E[w(\rho t - e_I(t) + e_I(0))]$$

and $A(t)$, $v(t)$, and $w(t)$ are defined in Chapter 5; graphs of the functions $u(t)$, $v(t)$, and $w(t)$ are given in Figure 5.18. The Fourier transform of the covariance is given by [7.1]:

$$\Phi_S(f) = \Phi_{S1}(f) + \Phi_{S2}(f) + \Phi_{S3}(f) \tag{7.12}$$

where

$$\Phi_{S1}(f) = F\{C_{S1}(t)\} = \text{sinc}^2(f)\, \tilde{Q}(f) \tag{7.13}$$

in which

$$\tilde{Q}(f) = \sum_{n=1}^{\infty} \left(\frac{1}{2\pi n}\right)^2 \left\{ \sum_{k=-\infty}^{\infty} Z_n(f - n\rho f_{j,max} - kf_{j,max}) \right.$$

$$\left. + \sum_{k=-\infty}^{\infty} Z_n(-f - n\rho f_{j,max} - kf_{j,max}) \right\} \tag{7.14}$$

$$Z_n(f) = F\{z_n(t)\} \tag{7.15}$$

$$z_n(t) = E[\exp(-j2\pi n (e_I(t) - e_I(0))] \tag{7.16}$$

$$\Phi_{S2}(f) = F\{C_{S2}(t)\} = \sum_{n=1}^{\infty} \left(\frac{\rho}{2\pi n}\right)^2 \{\delta(f - nf_{j,max}) + \delta(f + nf_{j,max})\} \tag{7.17}$$

and

$$\Phi_{S3}(f) = F\{C_{S3}(t)\} = \text{sinc}^2(f) \sum_{k=-\infty}^{\infty} \Phi_I(f - kf_{j,max}) \tag{7.18}$$

where frequency and time are once again expressed in traditional units. Therefore, the power spectral density is

$$\Phi_S(f) = \text{sinc}^2(f)\, \tilde{Q}(f) + \sum_{n=1}^{\infty} \left(\frac{\rho}{2\pi n}\right)^2 \cdot \{\delta(f - nf_{j,max})$$

$$+ \delta(f + nf_{j,max})\} + \text{sinc}^2(f) \sum_{k=-\infty}^{\infty} \Phi_I(f - kf_{j,max}) \tag{7.19}$$

Input jitter makes its presence felt in both the appearance of the feed-through term

$$\operatorname{sinc}^2(f) \sum_{k=-\infty}^{\infty} \Phi_I(f - kf_{j,max})$$

and in the smearing of the spectral lines. Note that $\tilde{Q}(f)$ has the same functional form as $Q(f)$, but the envelope functions $Z_n(f)$ replace impulses. These envelope functions are difficult to compute in general; however, exact evaluations for sinusoidal and Gaussian input jitter have been carried out [7.1].

The effect of sinusoidal input jitter on multiplex output jitter has been addressed in Chapter 6. The effect of Gaussian distributed input jitter, on multiplex output jitter, is addressed in this section. Consider as an example Gaussian distributed input jitter, with a power spectrum similar to that of cascaded T1 type line regenerators (see Figure 7.1 [7.2]). The amplitude distribution of the jitter, as illustrated, is approximately Gaussian for long chains of regenerators and the power spectrum is flat for low frequencies, falling off as the inverse square of frequency for higher frequencies.

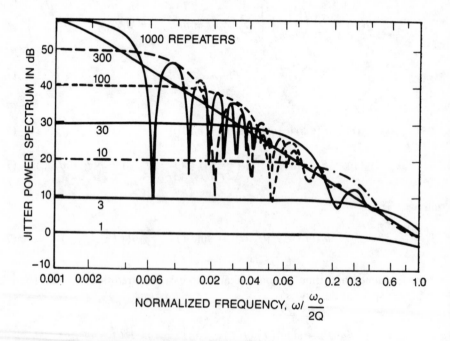

Fig. 7.1 T1 type regenerator jitter spectrum

Let σ_I represent the rms amplitude of the input jitter and B represent the corner frequency of the power spectrum. We then let

$$\Phi_I(f) = \frac{1}{2\pi} \sigma_I^2 \frac{2B}{B^2 + f^2} \tag{7.20}$$

approximate the power spectrum of the input jitter [7.1]. Finding the inverse Fourier transform of (7.20) gives

$$C_I(t) = \sigma_I^2 e^{-2\pi B|t|} \tag{7.21}$$

Because the input jitter is Gaussian, we can express $z_n(t)$ as [7.1]:

$$z_n(t) = e^{-(2\pi n)^2 C_I(0)} \sum_{k=0}^{\infty} \frac{(2\pi n)^{2k}}{k!} \{C_I(t)\}^k \tag{7.22}$$

Therefore

$$z_n(t) = e^{-(2\pi n\sigma_I)^2} \sum_{k=0}^{\infty} \frac{(2\pi n\sigma_I)^{2k}}{k!} e^{-2\pi kB|t|} \tag{7.23}$$

and

$$Z_n(f) = e^{-(2\pi n\sigma_I)^2} \sum_{k=0}^{\infty} \frac{(2\pi n\sigma_I)^{2k}}{\pi k!} \frac{kB}{(kB)^2 + f^2} \tag{7.24}$$

Thus

$$\begin{aligned}
\tilde{Q}(f) = \sum_{n=1}^{\infty} & \frac{e^{-(2\pi n\sigma_I)^2}}{(2\pi n)^2} \sum_{k=0}^{\infty} \frac{(2\pi n\sigma_I)^{2k}}{\pi k!} \\
& \cdot kB \sum_{q=-\infty}^{\infty} \left\{ \frac{1}{(kB)^2 + (f - qf_{j,max} - n\rho f_{j,max})^2} \right. \\
& \left. + \frac{1}{(kB)^2 + (f - qf_{j,max} + n\rho f_{j,max})^2} \right\}
\end{aligned} \tag{7.25}$$

The jitter power after filtering by the desynchronizer phase-smoothing circuit, P_D, may be obtained by integrating the filtered multiplex jitter spectrum from $-\infty$ to $+\infty$. Using (7.12)

$$P_D = \sigma_D^2 = \int_{-\infty}^{\infty} |G(f)|^2 \, \Phi_S(f) \, df = P_{D1} + P_{D2} + P_{D3} \qquad (7.26)$$

where $G(f)$ is the transfer function of the desynchronizer phase-smoothing circuit. Substituting (7.19) into (7.26) above, the filtered power at the demultiplexer output is

$$
\begin{aligned}
P_D &= \int_{-\infty}^{\infty} |G(f)|^2 \left\{ \mathrm{sinc}^2(f) \, \tilde{Q}(f) \right. \\
&\quad + \sum_{n=1}^{\infty} \left(\frac{\rho}{2\pi n} \right)^2 [\delta(f - nf_{j,max}) + \delta(f + nf_{j,max})] \\
&\quad \left. + \mathrm{sinc}^2(f) \sum_{k=-\infty}^{\infty} \Phi_I(f - kf_{j,max}) \right\} df \\
&= \int_{-\infty}^{\infty} |G(f)|^2 \, \mathrm{sinc}^2(f) \, \tilde{Q}(f) df + 2 \sum_{n=1}^{\infty} \left(\frac{\rho}{2\pi n} \right)^2 |G(nf_{j,max})|^2 \\
&\quad + \int_{-\infty}^{\infty} |G(f)|^2 \, \mathrm{sinc}^2(f) \cdot \sum_{k=-\infty}^{\infty} \Phi_I(f - kf_{j,max}) df \qquad (7.27)
\end{aligned}
$$

Note that the third term is a constant independent of ρ. For low frequencies

$$\mathrm{sinc}^2(f) \cdot \sum_{k=-\infty}^{\infty} \Phi_I(f - kf_{j,max}) \approx \Phi_I(f)$$

Therefore, the third term is the power of the jitter that would be present at the output of the multiplexer-demultiplexer pair, if its only effect were to filter the input jitter. Within this approximation, the jitter power added by bit justification is from the first and second terms

$$\Delta P_D = P_{D1} + P_{D2} \qquad (7.28)$$

Figure 7.2 [7.1] illustrates a theoretical graph of output jitter power, for a $DS1$ to $DS2$ multiplex with parameters as defined in Chapter 5, as a function of justification ratio for $\sigma_I = .1 \, DS1$ UI, and $B = 536.7$ Hz. The effect of the input jitter is to erode the peaks and valleys illustrated in Figure 5.16 for the same multiplex. Furthermore, it is expected that any type of input jitter would produce a similar erosion. Thus, the choice of justification ratio is not as critical as it appeared to be in Chapter 5.

The analysis which has been presented within this section may be applied to the case of negative justification. In addition, using the general principles developed above, expressions may also be derived for multiplex output jitter in the presence of input jitter for positive-negative and positive-zero-negative justification.

It is possible to estimate an upper bound for the accumulation of filtered multiplex jitter in cascaded bit justification multiplexes [7.1]. The terms P_{D1} and P_{D3} are

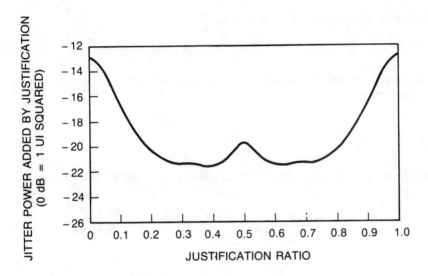

MULTIPLEX JITTER POWER AS A FUNCTION OF
JUSTIFICATION RATIO WITH GAUSSIAN INPUT JITTER

NOTES: INPUT GAUSSIAN JITTER OF RMS AMPLITUDE = .1 UI
RC SPECTRUM WITH A CORNER FREQUENCY = 536.7 Hz
DESYNCHRONIZED FILTER CORNER FREQUENCY = 644 Hz

Fig. 7.2 Multiplex jitter power as a function of justification ratio with Gaussian input jitter

difficult to evaluate in general, but the term P_{D2} can easily be evaluated if $G(f)$ is known. Generally, the desynchronizer phase-smoothing circuit bandwidth will be much smaller than the maximum justification rate; therefore, P_{D2} will be quite small in comparison with P_{D1} and P_{D3}.

To obtain a bound on the power of multiplex jitter, define

$$G_{max} = \max_{f\epsilon(-\infty,\infty)} |G(f)| \qquad (7.29)$$

We will first obtain a bound for P_{D1}. Using (7.12), (7.26), and (7.29)

$$P_{D1} \leq G_{max}^2 \int_{-\infty}^{\infty} \Phi_{S1}(f)df \qquad (7.30)$$

However,

$$\int_{-\infty}^{\infty} \Phi_{S1}(f)df$$

is the power, given by $C_{S1}(0)$. Therefore,

$$P_{D1} \leq G_{max}^2 C_{S1}(0)$$

Referring to (7.9) and the definitions of $A(t)$ and $w(t)$ given in Chapter 5, $C_{S1}(0) = 1/12$. Therefore

$$P_{D1} \leq \frac{1}{12} G_{max}^2 \qquad (7.31)$$

A bound on P_{D2} may easily be obtained. Using (7.27) and (7.29)

$$P_{D2} = 2 \sum_{n=1}^{\infty} \left(\frac{\rho}{2\pi n}\right)^2 |G(nf_{j,max})|^2 \leq 2G_{max}^2 \sum_{n=1}^{\infty} \left(\frac{\rho}{2\pi n}\right)^2$$

$$\leq \frac{G_{max}^2 \rho^2}{12} \qquad (7.32)$$

Finally, we will obtain a bound for P_{D3}. Again using (7.12), (7.26), and (7.29)

$$P_{D3} \leq G_{max}^2 \int_{-\infty}^{\infty} \Phi_{S3}(f)df$$

Using (7.18)

$$P_{D3} \leq G_{max}^2 C_{S3}(0)$$

Referring to (7.11), $C_{S3}(0) = C_I(0) = P_I$. Thus

$$P_{D3} \leq G_{max}^2 P_I \qquad (7.33)$$

Therefore, using (7.26), an upper bound on the filtered multiplex jitter power is

$$P_D \leq G_{max}^2 \left[\frac{1}{12} + \frac{\rho^2}{12} + P_I\right] \qquad (7.34)$$

and its corresponding rms amplitude bound is

$$\sigma_D = G_{max} \left[\frac{1}{12} + \frac{\rho^2}{12} + P_I \right]^{1/2} \qquad (7.35)$$

For a more conservative bound, as $\rho \leq 1$,

$$P_D \leq G_{max}^2 \left[\frac{1}{12} + \frac{1}{12} + P_I \right] = G_{max}^2 \left[\frac{1}{6} + P_I \right] \qquad (7.36)$$

This is a conservative estimate, because P_{D2} will actually be extremely small due to typical phase-smoothing circuit bandwidths. It is probably more realistic to use

$$P_D \leq G_{max}^2 \left(\frac{1}{12} + P_I \right)$$

If there is no peaking in $G(f)$, $G_{max}^2 = |G(0)|^2 = 1$ and

$$P_D \leq \frac{1}{12} + \frac{\rho^2}{12} + P_I \qquad (7.37)$$

Using this result, an upper bound for the accumulation of jitter power can be obtained for cascaded bit justification multiplexes [7.1].

Figure 7.3 illustrates N multiplexes in cascade. Denote the jitter on the output of the Kth synchronizer by $e_{SO,K}(t)$ and the jitter on the output of the Kth desynchronizer by $e_{DO,K}(t)$, and let $P_{SO,K}(f)$ and $P_{DO,K}(f)$ denote their respective powers. Neither the justification ratios at each of the synchronizers, nor the overall transfer functions of the phase-smoothing circuits in the N desynchronizers are assumed identical. Defining

$$G_{max,K} = \max_{f \in (-\infty, \infty)} |G_K(f)| \qquad (7.38)$$

Fig. 7.3 N multiplexes in cascade

then

$$P_{DO,K} \leq G_{max,K}^2 \left\{ \frac{1}{12} + \frac{\rho_K^2}{12} + P_{SI,K} \right\} \qquad (7.39)$$

where $P_{SI,K} = P_{DO,K-1}$. If the only input jitter is from the preceding multiplexer-demultiplexer pair, $P_{DO,0} = 0$ (no input jitter to the first multiplex). Thus

$$P_{DO,N} \leq \frac{1}{12} \sum_{K=1}^{N} \prod_{q=K}^{N} G_{max,q}^2 + \frac{1}{12} \sum_{K=1}^{N} \rho_K^2 \prod_{q=K}^{N} G_{max,q}^2 \qquad (7.40)$$

Defining

$$M = \max_{K=1,\ldots,N} G_{max,K}^2 \qquad (7.41)$$

and

$$\rho_M = \max_{K=1,\ldots,N} \rho_K \qquad (7.42)$$

then

$$P_{DO,N} \leq \frac{1}{12} (1 + \rho_M^2) \sum_{K=1}^{N} M^{N+1-K} \qquad (7.43)$$

If there is no peaking in any of the phase smoothing circuits, $M = 1$, and

$$P_{D,N} \leq \frac{N}{12} (1 + \rho_M^2) \qquad (7.44)$$

As $\rho_M \leq 1$, $P_{D,N} \leq N/6$. Therefore, the rms jitter amplitude at the output of the Nth multiplex is no greater than

$$\sigma = \sqrt{\frac{N}{6}} \qquad (7.45)$$

Empirical data [7.3] has indicated that the amplitude probability density function of filtered multiplex jitter is primarily a function of the multiplex justification ratio, and of the input jitter characteristics. If the justification ratio is exactly a ratio of small integers, the unfiltered multiplex inherent jitter waveform is periodic, and the tails of the density function truncate sharply. However, for other justification ratios the tails truncate less sharply. With Gaussian input jitter, the tails of the output jitter amplitude probability density function tend towards those of a Gaussian distribution with the same rms amplitude. Figures 7.4 a–c [7.3, 7.4] illustrate inherent DS1 and DS2 multiplex amplitude distribution statistics with justification ratios of 0.3333, 0.3346, and 0.3800, respectively. As expected, the amplitude distribution for $\rho = 0.3800$ truncates less rapidly than that for $\rho = 0.3333$.

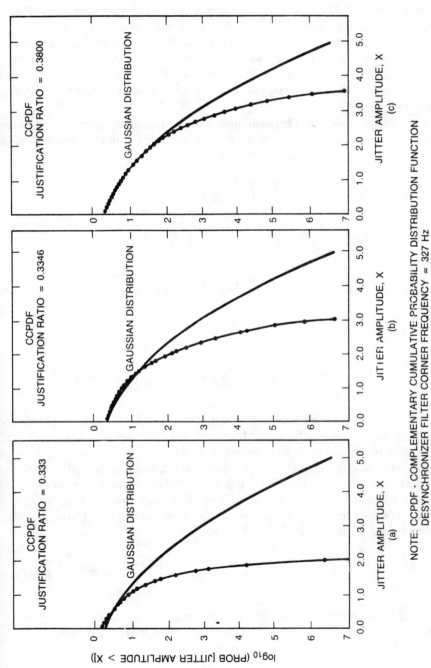

NOTE: CCPDF - COMPLEMENTARY CUMULATIVE PROBABILITY DISTRIBUTION FUNCTION
DESYNCHRONIZER FILTER CORNER FREQUENCY = 327 Hz

Fig. 7.4 Measured DS1 to DS2 multiplex jitter amplitude distribution statistics

On the basis of this data, a theoretical model for predicting jitter accumulation in cascaded bit justification multiplexes has been developed [7.5] which assumes that the jitter from each multiplex in the chain can be modeled by filtered Gaussian random variables. (The impact and accuracy of this assumption is discussed in more detail later within this section.) The jitter spectrum, $\Phi(f)$, and total power, σ^2, after each succeeding multiplex, is computed as the accumulation due to the preceding multiplexes.

Substituting into (7.19) Gaussian input jitter having an rms amplitude of σ_I and double-sided power spectral density $\Phi_I(f)$, the unfiltered multiplex output jitter is given by

$$
\begin{aligned}
\Phi_S(f) = \mathrm{sinc}^2(f) \sum_{n=1}^{\infty} \frac{1}{(2\pi n)^2} &\left(\sum_{k=-\infty}^{\infty} Z_n\left(f - n\rho f_{j,max} - kf_{j,max}\right) \right.\\
&\left. + Z_n\left(-f - n\rho f_{j,max} - kf_{j,max}\right) \right) + \sum_{n=1}^{\infty} \frac{\rho^2}{(2\pi n)^2} \left[\delta(f - nf_{j,max})\right. \\
&\left. + \delta(f + nf_{j,max})\right] + \mathrm{sinc}^2(f) \sum_{k=-\infty}^{\infty} \Phi_I\left(f - kf_{j,max}\right)
\end{aligned}
\tag{7.46}
$$

where [7.1]

$$
Z_n(f) = e^{-(2\pi n\sigma_I)} \left[\delta(f) + \sum_{k=1}^{\infty} \frac{(2\pi n)^{2k}}{k!} \underbrace{\Phi_I(f) * \ldots * \Phi_I(f)}_{k \text{ terms}} \right] \tag{7.47}
$$

The analysis up to this point allows estimation of the rms jitter amplitude for cascaded bit justification multiplexes. However, from a systems design standpoint, determination of peak-to-peak jitter amplitude is critical. Once again making use of the assumption that the amplitude distribution of jitter from cascaded multiplexes approaches a Gaussian, it is possible to calculate a "peak" value that is not exceeded more than a certain percent of the time. The probability that the jitter exceeds a particular threshold amplitude $|x|$ n times in the time interval $(t, t + \Delta t)$, has been described by a Poisson density function [7.5]:

$$
\Pr\{n(\pm x) \text{ crossings in } (t, t + \Delta t)\} = \frac{\{\overline{N(x)} \Delta t\}^n}{n!} e^{-\overline{N(x)}\Delta t} \tag{7.48}
$$

where $\overline{N(x)}$ is the average number of times per second that the threshold $|x|$ is exceeded. For Gaussian jitter, with power spectrum $\Phi(f)$, $\overline{N(x)}$ is given by

$$
\overline{N(x)} = N_0\, e^{-x^2/2\sigma^2} \tag{7.49}
$$

where

$$\sigma^2 = \int_{-\infty}^{\infty} \Phi(f)df \tag{7.50}$$

and

$$N_0 = 2 \left\{ \frac{\int_{-\infty}^{\infty} f^2 \Phi(f)df}{\sigma^2} \right\}^{1/2} \tag{7.51}$$

The condition that the jitter doesn't exceed a particular threshold *more* than n times during the time interval $(t, t + \Delta t)$ can be expressed as

$$\sum_{m=0}^{n} \frac{\{\overline{N(x)} \Delta t\}^m}{m!} e^{-\overline{N(x)}\Delta t} = 1 - P_n \tag{7.52}$$

where P_n is the probability of exceeding the threshold *more* than n times in the time interval $(t, t + \Delta t)$.

Then, the probability that the jitter doesn't exceed the threshold during the time interval $(t, t + \Delta t)$ is

$$1 - P_0 = e^{-\overline{N(x)}\Delta t} \tag{7.53}$$

Solving for the threshold

$$|x| = \left\{ 2\sigma^2 \ln \left[N_0 \frac{\Delta t}{\ln\left(\dfrac{1}{1 - P_o}\right)} \right] \right\}^{1/2} \tag{7.54}$$

If we assume that each time the threshold is crossed, an undesirable event (impairment) may result, the *mean time between impairments* (MTBI) may be taken as [7.6]

$$\text{MTBI} = \frac{1}{\overline{N(x)}} \tag{7.55}$$

Using (7.53), and solving for $\overline{N(x)}$

$$\overline{N(x)} = \frac{\ln\left(\dfrac{1}{1 - P_o}\right)}{\Delta t} \tag{7.56}$$

Therefore,

$$\text{MTBI} = \frac{\Delta t}{\ln\left(\dfrac{1}{1 - P_o}\right)} \tag{7.57}$$

Thus, (7.54) may be expressed as

$$|x| = \sqrt{2\sigma^2 \ln (N_0 \text{ MTBI})} \tag{7.58}$$

Note that using (7.54), an expression may be developed which relates the peak amplitude during one observation interval to the peak amplitude during a different observation interval. For example, if the output jitter is measured for a *convenient* specified observation time interval (e.g., one minute) and the amplitude statistics of the jittered signal are known, then it is possible to make a prediction concerning the peak jitter amplitude if a longer observation interval were utilized (e.g. one day). Defining $|x_1|$ as the magnitude of the maximum peak amplitude which occurred during observation interval ΔT_1, $|x_2|$ as the magnitude of the maximum peak amplitude which should occur during interval ΔT_2, and σ as the rms jitter amplitude

$$|x_2| = \sqrt{|x_1|^2 + 2\sigma^2 \ln \left(\frac{\Delta T_2}{\Delta T_1}\right)} \tag{7.59}$$

As discussed earlier, the analysis of jitter from cascaded bit justification multiplexes using first order frequency domain techniques relies on an assumption of Gaussian input jitter. Time domain simulation techniques (discussed in Chapter 5, and further below), may be utilized in order to determine a first-order distribution characterization of multiplex jitter [7.7], and further assess the validity of this assumption.

In Chapter 5, a time domain simulation approach was described which allows calculation of inherent multiplex jitter. This technique may be generalized to handle input jitter, thereby generating a sampled time waveform of multiplex jitter for arbitrary input jitter, and may therefore be utilized to compute multiplex jitter accumulation.

If the incoming signal is jittered, the approach used to derive (5.40) is still valid except that the period of the jittered incoming lower rate signal, denoted T, is not constant. Then,

$$\Delta_1 e_{\text{SPC}}[nT] = T - (1 + n_{oh})T_r \tag{7.60}$$

The relationship between T_{nom} and T may be derived by considering two successive bits which would ideally appear at times nT_{nom} and $(n + 1)T_{nom}$, but are jittered by $e_I[nT_{nom}]$ and $e_I[(n + 1)T_{nom}]$, respectively. Referring to Figure 7.5, we see that

$$e_I[nT_{nom}] + T_{nom} = T + e_I[(n + 1)T_{nom}]$$

Solving for T,

$$T = T_{nom} - (e_I[(n + 1)T_{nom}] - e_I[nT_{nom}])$$
$$= T_{nom} - \Delta_1 e_I[nT_{nom}] \tag{7.61}$$

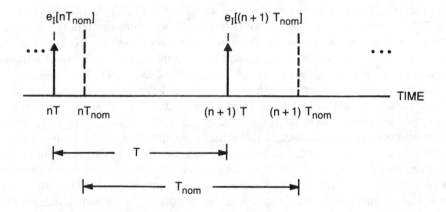

Fig. 7.5 Two successive jittered impulses

Therefore,

$$\Delta_1 e_{\text{SPC}}[nT] = T_{nom} - \Delta_1 e_I[nT_{nom}] - (1 + n_{oh})T_r \tag{7.62}$$

Substituting (5.40) for $e_S[nT_{nom}]$ into (7.62) above, we have

$$\Delta_1 e_{\text{SPC}}[nT] = \Delta_1 e_S[nT_{nom}] - \Delta_1 e_I[nT_{nom}] \tag{7.63}$$

which is easily rearranged into the format of (7.1).

Using the techniques discussed above to simulate the synchronizer mechanism with input jitter, an appropriate desynchronizer filter transfer characteristic should be chosen so that the filtered multiplex output jitter may be determined. In a similar fashion as for the first order frequency domain approach, the output jitter after each multiplex may then be computed as the accumulation due to the preceding multiplexes. Figure 7.6 [7.8] illustrates an example of a simulator flow chart for computing jitter accumulation from cascaded bit justification multiplexes.

Multiplex jitter has been modeled as a random process whose first order distribution belongs to the generalized Gaussian family [7.7], whose density function is given by

$$f(x) = B \exp\left\{ -\frac{b}{\sigma^2} |x - \bar{x}|^c \right\} \tag{7.64}$$

where the constants B and b are determined by the expression for the variance and the fact that the density must integrate to unity. The generalized Gaussian family was chosen because it exhibits the following desirable properties: it contains the Gaussian distribution ($c = 2$ in Figure 7.7), and it encompasses a wide range of distribution tail behavior. A representative sample of the family is illustrated in Figure 7.7. In order to select an appropriate distribution from the family, maximum likelihood and method-of-moment estimators were developed for the variables that parameterize the family, and random samples from the time-domain jitter simulation were used as the data for the estimators. For cascaded DS1 to DS2 multiplexes with randomized justification ratios, with the format and desynchronizer phase smoothing circuit parameters defined in Chapter 5, it was found that while the estimators of the family parameters were far from those of the Gaussian after a single multiplex, the estimators rapidly approached the values corresponding to the Gaussian distribution. In addition, quantile-quantile plots[1] against the Gaussian distribution also indicated that the empirical density of the cascaded DS1 to DS2 multiplex jitter was close to the Gaussian for multiplex chain lengths greater than four. Thus, the results provide confirmation of the assumption that jitter from cascaded bit justification multiplexes becomes Gaussian.

7.2 JITTER ACCUMULATION IN EXISTING DIGITAL NETWORKS

A frequency domain based model of the network jitter accumulation has been developed [7.4] which assumes that the jitter from each of the component parts in the network i.e., cascaded line regenerators and bit justification multiplexes, can be modeled by filtered Gaussian random variables.

The power spectrum and rms amplitude after each component part is computed as the accumulation due to the preceding parts according to the following rules:

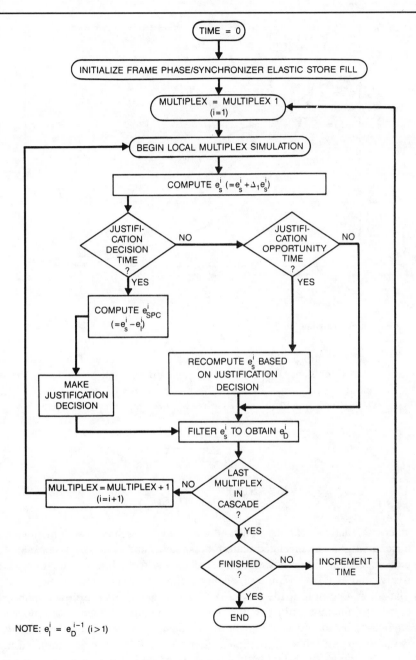

NOTE: $e_I^i = e_D^{i-1}$ (i > 1)

Fig. 7.6 Example of simulation of jitter accumulation from cascaded multiplexes

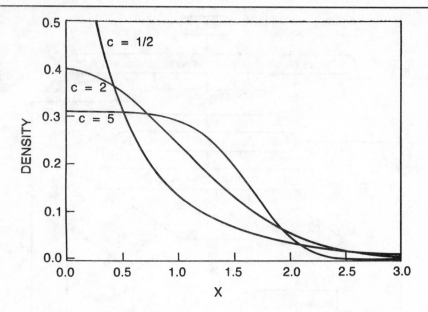

Fig. 7.7 Generalized Gaussian family

1. The jitter spectrum at the output of a chain of line regenerators is the power sum of the jitter generated by the line regenerators and any jitter at the input of the chain, appropriately filtered by the equivalent jitter transfer characteristic. Thus, for input jitter, $\Phi_I(f)$, the output jitter, $\Phi(f)$, is given by

$$\Phi(f) = \Phi_i^R |H(f)|^2 \frac{1 - |H(f)|^{2N}}{1 - |H(f)|^2}$$

$$+ \Phi_i^S |H(f)|^2 \frac{|1 - H(f)^N|^2}{|1 - H(f)|^2} + \Phi_I(f)|H(f)^N|^2 \qquad (7.65)$$

where Φ_i^R and Φ_i^S are the constant, internally generated, random (pattern independent plus uncorrelated pattern dependent) and systematic (correlated pattern dependent) jitter power spectral densities for a line regenerator, respectively, and $H(f)$ is the line regenerator jitter transfer characteristic.

2. The jitter spectrum at the output of a demultiplexer is the power sum of the unfiltered inherent multiplex jitter and the accumulated higher rate input jitter, attenuated by the desynchronizer jitter transfer characteristic. Thus, if $\Phi_{I1}(f)$ is the unfiltered inherent multiplex jitter and $\Phi_{I2}(f)$ is the accumulated higher rate input jitter, the output jitter, $\Phi(f)$, is given by

$$\Phi(f) = \left\{ \Phi_{I1}(f) + \frac{\Phi_{I2}(f)}{r^2} \right\} |G(f)|^2 \qquad (7.66)$$

where r is the multiplexing ratio and $G(f)$ represents the desynchronizer jitter transfer characteristic.

The threshold crossing theory, utilized to predict peak multiplex jitter amplitudes is also applicable for predicting peak network jitter amplitudes.

7.3 JITTER IMPLICATIONS OF COUPLING SONET AND EXISTING DIGITAL NETWORKS

As discussed in Chapter 5, pointer adjustments generate no jitter on synchronous optical signals. However, they do impact the jitter characteristics of existing standard digital signals multiplexed into synchronous optical signals. Consider as an example the scenario illustrated within Figure 7.8 of an asynchronous DS3 (44.736 Mb/s) signal traversing a hybrid network comprised of a synchronous optical network connecting two existing digital networks. At the interface between the existing and synchronous optical network, a "gateway" SONET multiplexer rate adapts the incoming signal using bit justification techniques,[2] and adds to this Synchronous Payload Envelope overhead common to the STS-1 signal which includes the pointer. This synchronous optical signal now traverses the synchronous optical network where it may be transported over repeatered fiber optic facilities and/or pass through various synchronous optical network elements which perform pointer processing. At the interface between the synchronous optical and existing networks the SONET "gateway" equipment demultiplexes the synchronous optical signal into its component DS3 tributaries. Within the SONET gateway demultiplexer, the received synchronous optical signal (generally, STS-N) clock is divided by N, and is used to clock the resulting STS-1 signals into associated desynchronizers. Within a desynchronizer, this clock is inhibited to account for incoming (positive) pointer adjustments, STS-1 frame overhead bytes, as well as STS-1 Synchronous Payload Envelope overhead bytes and justification bits. The resulting gapped clock must be smoothed by some mechanism to reduce jitter on the outgoing signal. In addition, this clock may have jitter on it from intervening transmission facilities, and clock and environmental sources. Most of the above jitter sources have already been considered in this and other chapters. However, two of these sources will be considered further as they require additional analysis:

- The effect of the pointer process on very low frequency synchronous optical network jitter (wander).
- The impact of the eight unit interval phase steps caused by pointer adjustments.

It is important to examine the impact of the pointer process on synchronous optical network wander to determine whether significant differences in wander at the hybrid network interface may result from the pointer processing mechanism. A simple mathematical model has been developed [7.9, 7.10] which predicts what effect the interaction of pointer processing and timing signal wander will have on the wander characteristics of a payload that has been transported through a chain of syn-

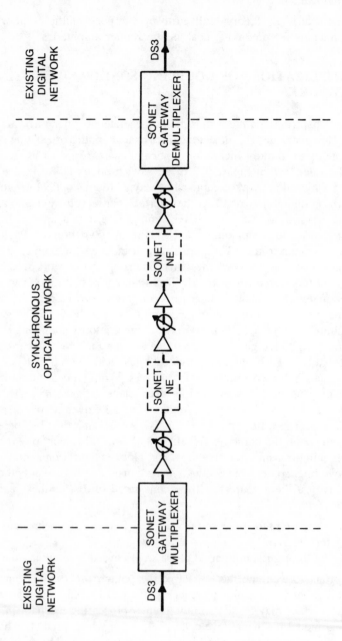

Fig. 7.8 Hybrid network: example scenario

chronous optical network elements. Figure 7.9 illustrates a pointer processing elastic store for the nth synchronous optical network element with STS-1 inputs. The wander affecting the phase of the write clock (relative to an ideal STS-1 clock) is denoted $\Phi_{in,n}$, and the wander affecting the phase of the read clock (relative to an ideal STS-1 clock) is denoted $\Phi_{out,n}$. Then, using these definitions, the relative wander (phase difference) between the read and write clocks within the elastic store is given by

$$\Delta\Phi = \Phi_{in,n} - \Phi_{out,n} \qquad (7.67)$$

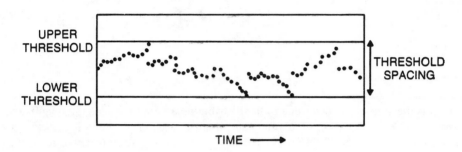

Fig. 7.9 Pointer processing elastic store

The first simplifying assumption made is that for "large" differential wander (i.e. peak-to-peak $\Phi_{in,n} - \Phi_{out,n} \gg$ upper and lower pointer adjustment threshold spacing in the pointer processing elastic store), the dead zone or guard-band effect of the threshold spacing[3] may be neglected (see Figure 7.9). In this case, the pointer processing mechanism quantizes the phase difference to a resolution of one byte (by definition). If the quantized signal x is represented by $Q[x]$ then the phase of the outgoing pointer adjustment, $P_{out,n}$, may be defined as

$$P_{out,n} \equiv Q[\Delta\Phi] = Q[\Phi_{in,n} - \Phi_{out,n}] \qquad (7.68)$$

If the quantization process is assumed to be approximately distributive with some error, ε_d, then $P_{out,n}$ may be represented as

$$P_{out,n} = Q[\Phi_{in,n}] - Q[\Phi_{out,n}] + \varepsilon_d \qquad (7.69)$$

Then the wander on the write clock signal into the $n + 1th$ SONET STS-1 pointer processing elastic store relative to an ideal STS-1 clock will be given by

$$\Phi_{in,n+1} = P_{out,n} + \Phi_{out,n} \qquad (7.70)$$

Substituting the approximate expression for $P_{out,n}$ into (7.70) above

$$\Phi_{in,n+1} = Q[\Phi_{in,n}] - Q[\Phi_{out,n}] + \Phi_{out,n} + \varepsilon_d \qquad (7.71)$$

Approximating

$$\Phi_{out,n} = Q[\Phi_{out,n}] + \varepsilon_{q,out} \qquad (7.72)$$

then

$$\Phi_{in,n+1} = Q[\Phi_{in,n}] + \varepsilon \qquad (7.73)$$

where

$$\varepsilon = \varepsilon_d + \varepsilon_{q,out} \qquad (7.74)$$

which is the sum of the error due to the distributive assumption plus the quantization error from $\Phi_{out,n}$. Equation (7.73) indicates that the wander on the write clock signal into the $n + 1th$ SONET STS-1 pointer processing elastic store (relative to an ideal STS-1 clock) is equivalent to a quantized version of the wander on the write clock signal into the nth SONET STS-1 pointer processing elastic store (relative to an ideal STS-1 clock) plus a noise term.

If the peak-to-peak amplitude of $\Delta\Phi$ is less than the threshold spacing (i.e. the assumption of $\Delta\Phi \gg$ threshold spacing is not valid), then

$$\Phi_{in,n+1} = \Phi_{out,n} \qquad (7.75)$$

The simplified mathematical model indicates that in general the wander on the write clock into the $n + 1th$ pointer processing elastic store is equivalent to a quantized version of the wander on the write clock into the nth pointer processing elastic store. As $\Phi_{in,n} = \Phi_{out,n}$ at the originating $(n = 0)$ pointer processing elastic store, by definition, $\Phi_{out,n}$ at the nth pointer processing store will be strongly correlated to the originating node wander.

Time domain simulation of the pointer process, to evaulate its impact on both random and deterministic sources of wander arising on transmission and synchronization distribution paths, has been performed [7.9, 7.10]. The simulation methodology is based upon phase accounting between the read and write clocks within the pointer processing store of the synchronous optical network element. In order to simplify the simulation, the waiting time between pointer adjustment request and pointer adjustment opportunity is ignored. Figure 7.10 [7.10] illustrates a simulator flow chart for single and cascaded pointer processing synchronous optical network elements. Simulation results agree with the predictions of the simplified mathematical model, indicating that the residual pointer adjustments at the end of a chain of

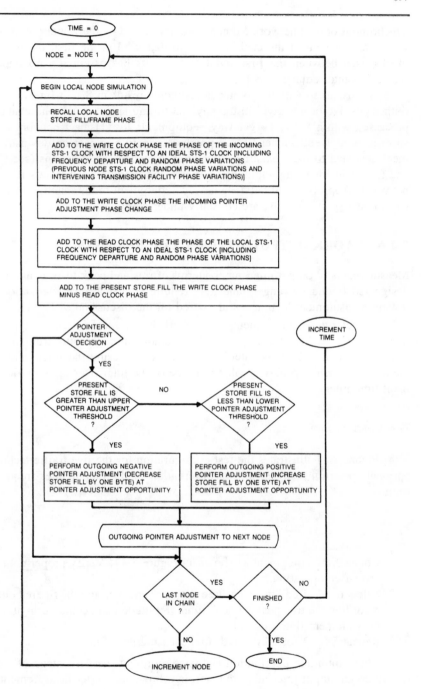

Fig. 7.10 Example of simulation of phase variation from single and cascaded pointer processing SONET network elements

synchronous optical network elements are dominated by the wander of the clock used in the last element of the chain relative to that used in the first; the wander on the clocks used between the first and last elements has essentially no impact on the residual pointer adjustments [7.9, 7.10].

As discussed earlier, pointer adjustments result in 8 unit interval phase steps (either positive or negative) in the payload timing signal. Such phase steps must be processed within SONET "gateway" equipments to: (1) meet network output jitter specifications (see Chapter 9), (2) preclude negative impact on downstream equipment designed to meet existing input jitter tolerance specifications (see Chapter 9), and (3) avoid imposing additional network configuration constraints in such hybrid networks. Appropriate SONET "gateway" equipment requirements, intended to satisfy the above constraints, are currently under study.

7.4 NETWORK OUTPUT JITTER MEASUREMENTS

Measurements of output jitter may be terms of rms and peak-to-peak amplitudes over designated frequency ranges, and may require statistical characterization. The following measurement techniques are based on the discussion of [7.11].

Output jitter measurements at network hierarchical interfaces typically use a live traffic signal. This technique involves demodulating the jitter from the live traffic at the output of a network interface, selectively filtering the jitter, and measuring the true rms or true peak-to-peak amplitude of the jitter over the specified measurement time interval.

Test Configuration

Figure 7.11 illustrates the test configuration for the live traffic technique. The optional spectrum analyzer allows observation of the output jitter frequency spectrum.

Procedure

1. Connect the equipment as shown in Figure 7.11. Verify proper continuity and error-free operation.
2. Select the desired jitter measurement filter and measure the filtered output jitter, recording the true peak-to-peak jitter amplitude that occurs during the specified measurement time interval.
3. Repeat Step 2 for all desired jitter measurement filters.

Jitter amplitude distribution statistics represent the percentage of total events for which an output jitter amplitude (from a broadband jitter measurement) exceeds a selected threshold amplitude. These statistics may be plotted as a function of the threshold jitter amplitude to provide the *complementary cumulative probability dis-*

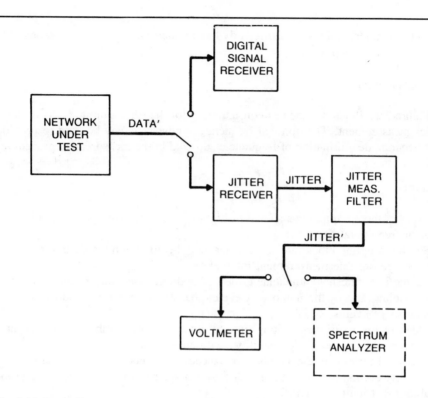

Fig. 7.11 Output jitter measurement configuration

tribution function (CCPDF). The CCPDF may be analyzed and compared with known distribution functions to ensure satisfactory equipment and network performance in the presence of jitter.

Jitter amplitude distribution statistics are measured by setting a threshold jitter amplitude and determining the proportion of time that jitter amplitudes larger than this threshold occur on the received signal. The process is repeated for a sufficient number of thresholds in order to accurately represent the continuous jitter probability distribution of the equipment, or network under test, over the applicable jitter amplitude range.

This technique utilizes digital processing to determine the proportion of time during which a jitter amplitude, larger than a selected threshold amplitude, occurs on the received signal. Specifically, the received jitter amplitude is compared to a selected threshold jitter amplitude, and sampled by a digital clock to obtain the number of events exceeding the threshold. A counter then displays the ratio of the exceeding events to the total number of possible events. A divider is used to scale the sampling clock, and to prevent the counter from overranging. The sampling clock frequency must be sufficiently high to provide for proper sampling of short duration exceeding events. The total number of events (which is dependent upon the measurement time interval) must be sufficiently large to determine the desired probability with a high confidence level.

Test Configuration

Figure 7.12 illustrates the test configuration for the jitter amplitude distribution statistics measurement. The optional frequency synthesizer may be used to provide a more accurate determination of frequencies utilized in the measurement procedure.

Procedure

1. Connect the equipment as shown in Figure 7.12. Verify proper continuity and error-free operation.
2. Set the sampling clock rate and divider modulus to obtain the desired accuracy over a reasonable measurement interval.
3. Set the jitter threshold amplitude as desired and clear the counter. Set the counter to display ratio of the number of events that exceed the threshold to the total number of events.
4. Measure for a sufficiently long time to obtain a large number of total events, and achieve the desired probability confidence level.
5. Record the ratio of the number of exceeding events to the total number of possible events, and repeat Steps 3–5 to characterize the jitter amplitude probability distribution (CCPDF).

7.5 REMARKS

In this chapter we have examined and characterized jitter accumulation from bit justification multiplexes, and discussed approaches for determining rms and peak-to-peak jitter amplitudes for cascaded bit justification multiplexes, and for bounding jitter accumulation. A model is presented which allows prediction of rms and peak-to-peak jitter amplitudes in networks comprised of line regenerators and bit justification multiplexes. In addition, the jitter implications of coupling synchronous optical and existing digital networks are explored. Finally, measurement techniques for determining and characterizing network jitter, are included.

NOTES

1. A familiar way of judging the fit of a Gaussian distribution to independent, identically distributed sampled data is to plot the empirical quantile points against the theoretical quantile points for the standard normal. Note that the 25% quantile point of the real random variable x is the real number a such that $P[x \leq a] = 0.25$.

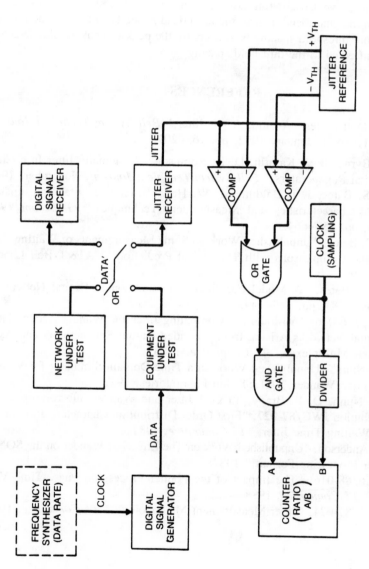

Fig. 7.12 Jitter amplitude distribution statistics measurement configuration

2. Overhead necessary for the end-to-end integrity of the incoming signal and fixed stuffing are also inserted.

3. A minimum threshold spacing of slightly more than one byte, a guardband, is required to prevent oscillating pointer adjustments. The system won't react to wander in the guardband; hence, the term dead zone. The effect of the nonzero elastic store threshold spacing is to decrease the peak-to-peak amplitude $\Delta\phi$ by an amount equal to the threshold spacing.

REFERENCES

[7.1] D. L. Duttweiller, "Waiting Time Jitter," *Bell System Technical Journal*, Vol. 51, No. 1, January 1972, pp. 165–207.

[7.2] C. J. Byrne, B. J. Karafin, D. B. Robinson, "Systematic Jitter In a Chain of Digital Regenerators," *Bell System Technical Journal*, November, 1963.

[7.3] R. J. S. Bates, J. J. Baldini, M. W. Hall, "Jitter in Digital Transmission Systems-Characteristics and Measurement Techniques," *Globecom 1982*, Miami, pp. 658–664.

[7.4] R. J. S. Bates, "Unpublished Work on Some Measurements of Waiting Time Jitter and Comparisons with Theoretical Predictions," AT&T Bell Laboratories, 1981.

[7.5] R. J. S. Bates, "A Model For Jitter Accumulation in Digital Networks," *Globecom '83*, pp. 145–149.

[7.6] T1X1.3/86-026, "A Model For Computing Jitter and Wander Accumulation in Digital Networks Arising from Cascaded Digital Regenerators and Asynchronous Multiplexes," *T1 Committee*, 1986.

[7.7] R. O. Nunn, "Unpublished Work on a Time-Domain Simulator for Waiting Time Jitter/Wander," AT&T Bell Laboratories, 1987.

[7.8] R. O. Nunn, D. P. Brady, T1X1.3 Jitter and Wander Subworking Group Contribution JWC/87-027, "First Order Distribution Characterization of Filtered Waiting Time Jitter," *T1 Committee*, 1987.

[7.9] R. D. Anderson, "Unpublished Work on The Effects of Wander on the SONET Pointer Processing Format," 1987.

[7.10] T1X1.6/88-016, "The Impact of the Pointer Process on Phase-Time Variation," *T1 Committee*, 1988.

[7.11] T1X1.4/87-024, "Jitter Measurement Methodology," *T1 Committee*, 1987.

Chapter 8
Wander
by Karen E. Plonty

In the previous chapters, we have discussed jitter introduced by line regenerators and digital multiplexes in detail. This chapter will focus on *wander*, an extremely slow pulse position modulation of the pulse stream, often varying with a period of one day (*diurnal wander*) or one year (*annual wander*). The study of wander departs from that of jitter mainly because the sources of wander, which are largely environmental. Slowly-varying environmental changes result in a "wandering" of the received pulse stream. Several potential sources of wander exist in a fiber optic transmission system. For example, changes in the fiber temperature, drift in the regenerator laser wavelength, and instabilities in the timing reference all lead to wander.

Except for waiting time wander, discussed in Chapter 5, wander in fiber optic transmission systems is generally the result of variations in the light propagation properties of the fiber itself. Such wander cannot be reduced in a fiber optic network, as it is the result of uncontrollable changes in the cable environment. The most that can be done to reduce the effects of temperature variation, for example, is to bury the cable at a depth of several feet. This reduces the daily temperature variation to a fraction of a degree, although larger temperature swings will be seen annually. Aerial cable, although more economical to install, will be subject to larger temperature variations both daily and annually, resulting in larger amounts of wander.

Although wander itself cannot be eliminated, the number of transmission impairments caused by wander can be reduced by increasing the size of elastic stores in the multiplex or other transmission equipment. This will be of particular importance in synchronous high-speed fiber-optic networks, which may require an elastic store operating at the line rate to control wander.

Problems caused by wander arise in a transmission network when data is multiplexed, demultiplexed, or switched between lines. Wander causes the bits to be displaced from their expected positions in time. Consequently, the incoming bits cannot be accurately identified, resulting in bit errors at the receiving terminal. One way to avoid such errors is to use an elastic store at the interface between the transmission line and the terminal equipment to resynchronize the data (see Figure 8.1).

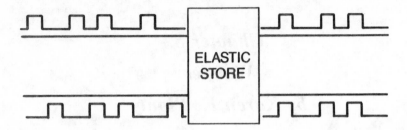

Fig. 8.1 Synchronization with an elastic store

The data is written into the elastic store by a clock that has been extracted from the received signal (see Chapter 2). Thus, the *phase-locked-loop* (PLL), or *surface acoustic wave* (SAW) filter used in the clock extraction circuit tracks the wander on the received signal. Bits are temporarily stored in the elastic store, then read out by a reference clock. By using the same reference clock to read the elastic stores, different data streams are forced into synchronization.

The elastic store of a transmission system must be large enough to cope with the variation in the time of arrival of the pulses in a transmission path relative to the other paths. When the accumulated wander exceeds the storage capacity of the elastic store, it overflows, resulting in a slip. Slips may cause error bursts and loss of frame synchronization. Although CCITT (*International Telegraph and Telephone Consultative Committee*) uncontrolled slip rate standards have not yet been set [8.1], it is not likely that many slips per day will be tolerable, so diurnal wander which causes slips must be controlled. Therefore, the size of the elastic store must be large enough to cope with all possible sources of variation in propagation delays that occur on a daily basis, including wander. Since annual wander is generally much larger than diurnal wander, it is impractical to use an elastic store with enough capacity to prevent slips on an annual basis. Such slips, therefore, should be considered as inevitable when setting future uncontrolled slip rate standards.

In this chapter we will model wander in fiber optic transmission systems and derive a simple model for wander accumulation. A wander measurement technique will be described and measurement results will be given.

8.1 WANDER IN FIBER OPTIC TRANSMISSION SYSTEMS

In an ideal fiber optic transmission system, the rising edge of the nth bit of a data stream occurs exactly at time nT. In real systems, however, as discussed in this book, the bits will deviate from their expected positions in time. In other words, the phase of each bit varies relative to the ideal system. Wander is defined as phase variations that occur with frequencies less than about 10 Hz. Diurnal wander, with a period of one day, and annual wander, with a period of one year, are particularly important because the causes of wander are largely environmental in nature.

Given that the propagation delay of an optical signal transmitted through an optical fiber of length l is [8.2]:

$$\tau = \frac{ln_c}{c} \tag{8.1}$$

where n_c is the group index of refraction of the fiber core, and we define wander as a slow change in τ with time, so that

$$w[nT] = \frac{d\tau}{dt}\bigg|_{t=nT} \tag{8.2}$$

Examining (8.1) and (8.2), we see that a slow change in l or n_c results in wander. Both optical fiber temperature changes and changes in laser wavelength will change n_c, while an optical fiber temperature change also results in a slight variation in the length of the optical fiber. Differentiating (8.1) with respect to temperature and wavelength, we find that

$$\frac{d\tau}{dt} = \frac{\partial \tau}{\partial \theta}\frac{d\theta}{dt} + \frac{\partial \tau}{\partial \lambda}\frac{d\lambda}{dt} \tag{8.3}$$

where θ is the fiber temperature and λ is the wavelength of the light propagating in the fiber. We will examine each term separately.

Fluctuation in optical fiber temperature is the most significant source of wander. Even a slight variation in temperature of the optical fiber can cause a significant amount of wander over a long distance, since both the group index of refraction and the length of the fiber are temperature-dependent. Using (8.2) and (8.3), the wander caused by variations in the optical fiber temperature in the Nth regenerator-fiber span is given by

$$w_{iN}^\theta[nT] = \frac{l}{c}\left[\frac{\partial n_c}{\partial \theta}\frac{d\theta}{dt}\right] + \frac{n_c}{c}\left[\frac{dl}{d\theta}\frac{d\theta}{dt}\right]\bigg|_{t=nT} \tag{8.4}$$

Fluctuation in the laser transmitter wavelength is another source of wander in fiber optic systems. Wavelength variations cause wander because the group refractive index of the optical fiber is wavelength-dependent. If the drift in laser wavelength is due entirely to changes in laser temperature, then

$$\frac{d\lambda}{dt} = \frac{d\lambda}{d\theta_l} \frac{d\theta_l}{dt}$$

where θ_l is the laser temperature. Substituting this result into (8.2) and (8.3), the wander caused by wavelength drift at the Nth regenerator-fiber span is

$$w_{iN}^{\lambda}[nT] = -\frac{l}{c} \frac{\partial n_c}{\partial \lambda} \frac{d\lambda}{d\theta_l} \frac{d\theta_l}{dt} \bigg|_{t=nT} \tag{8.5}$$

An additional source of wander is variation in the reference timing signal used to synchronize the clocks that control transmission terminals and switches. The transmission terminals and the switches in a synchronous network must operate at the same frequency with a high degree of stability (typically better than one part in 10^{11} [8.3]). However, stand-alone clocks that meet this stability requirement are expensive. Therefore, a reference timing signal may be distributed from a highly stable primary reference source at a central location to the lower level clocks through a synchronization reference distribution system.

In synchronous high-speed fiber optic networks, a reference timing signal at the high-speed line rate may be required. If the present master-slave synchronization architecture [8.1] is retained at high bit rates, a high-speed reference would probably be distributed over fiber optic transmission systems from a single primary reference source, or master clock. The primary reference source may have its own intrinsic wander. Furthermore, the fiber optic transmission systems used for synchronization reference distribution would be subject to the sources of wander previously discussed in this chapter.

Because wander frequencies may be too low to be filtered by typical PLL or SAW filters, it would be tracked by the lower level clocks receiving timing reference. These lower level clocks, in turn, could contribute their own intrinsic wander. As synchronization reference is distributed from the master clock through intermediate clocks and distribution facilities, the phase stability of the reference continues to degrade (i.e., wander accumulates unbounded in the synchronization reference distribution paths). The wander on the output of a clock slaved to a primary reference source is [8.4]

$$w^{f_0}[nT] = nTy_b + \frac{D_0}{2}(nT)^2 + e_{pm}[nT] + \int_{-\infty}^{nT} e_{fm}(t)\,dt \tag{8.6}$$

where y_b is a frequency offset that can arise from a disruption of the reference, D_0 is a linear frequency drift component, $e_{pm}(t)$ is white phase noise of the clock, and

$e_{fm}(t)$ is white noise frequency modulation of the clock because of a disrupted reference.

8.2 WANDER ACCUMULATION

In a long distance network, several sources of wander may be present along one transmission line. For example, each line incorporates many regenerators, each with a laser that may be a source of wander. The study of wander accumulation is greatly simplified because the jitter transfer function of fiber optic regenerators discussed in Chapter 2 approaches unity for jitter frequencies below a few kHz. Furthermore, jitter peaking is nonexistent at the low frequencies of interest when considering wander. Thus, the wander transfer function is unity and "wander peaking" need not be considered.

If the wander components from each source are systematic, the accumulated wander at the Nth regenerator-fiber span is

$$w_N^S[nT] = w_{i1}^S[nT] + w_{i2}^S[nT] + \ldots + w_{iN}^S[nT] \qquad (8.7)$$

where $w_{il}^S[nT]$ is the systematic component of wander at the lth regenerator-fiber span ($l = 1, 2, \ldots N$). The rms of the accumulated systematic wander is

$$\sigma_S[N] = \sigma_{i1}^S + \sigma_{i2}^S + \ldots + \sigma_{iN}^S \qquad (8.8)$$

where σ_{il}^S is the rms of the systematic wander from the lth regenerator-fiber span given by

$$\sigma_{il}^S = \sqrt{E\{|w_{il}^S[nT]|^2\}} \qquad (8.9)$$

Since variations in optical fiber temperature generally cover a wide geographic region, it can be assumed that $w_{iN}^\theta[nT]$ accumulates systematically according to (8.8).

If the wander components are random (i.e., independent of one another), the total amount of wander accumulated is

$$w_N^R[nT] = w_{i1}^R[nT] + w_{i2}^R[nT] + \ldots + w_{iN}^R[nT] \qquad (8.10)$$

where $w_{il}^R[nT]$ is the random component of wander at the lth regenerator-fiber span ($l = 1, 2, \ldots N$). The rms of the accumulated random wander is

$$\sigma_R[N] = \sqrt{(\sigma_{i1}^R)^2 + (\sigma_{i2}^R)^2 + \ldots + (\sigma_{iN}^R)^2} \qquad (8.11)$$

where σ_{il}^R is the rms of the random wander from the lth regenerator-fiber span given by

$$\sigma_{il}^R = \sqrt{E\{|w_{il}^R[nT]|^2\}} \tag{8.12}$$

Since each laser transmitter is independently temperature-controlled, $w_{iN}^\lambda[nT]$ accumulates randomly according to (8.11).

From (8.8) and (8.11), it can be observed that the rms of the systematic wander accumulates more rapidly than the rms of the random wander of the same magnitude. Since the sources of wander discussed in this chapter are independent of one another, it can further be stated that

$$w_N[nT] = w_N^R[nT] + w_N^S[nT] = w_N^\theta[nT] + w_N^\lambda[nT] + w^{f_0}[nT] \tag{8.13}$$

By making some realistic assumptions about the optical fiber properties, the magnitude of the optical fiber temperature change, the laser characteristics, and the variation in temperature, an estimate of the wander in an optical fiber transmission network can be made. Table 8.1 lists typical optical fiber properties and laser characteristics.

Table 8.1 Typical Optical Fiber and Laser Characteristics

Optical Fiber or Laser Characteristics	*Approximate Value*
$\dfrac{dn_c}{d\theta}$	1.2×10^{-5} per °C*
$\dfrac{1}{l}\dfrac{dl}{d\theta}$	8.0×10^{-7} per °C**
$\dfrac{1}{c}\dfrac{dn_c}{d\lambda}$	$l\,\dfrac{\text{ps}}{\text{nm km}}$ at 1.3μm
	$17\,\dfrac{\text{ps}}{\text{nm km}}$ at 1.5μm
$\dfrac{d\lambda}{d\theta_l}$	0.1 nm per °C

*[8.5]
**[8.6]

Data on soil temperature at a depth of about three feet (the nominal depth at which optical fiber cable is buried), illustrated in Figure 8.2, shows that the daily variation in soil temperature rarely exceeds 1°C, while the annual variation in soil temperature is about 20°C [8.7].

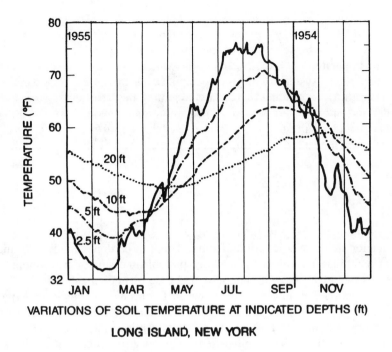

VARIATIONS OF SOIL TEMPERATURE AT INDICATED DEPTHS (ft)

LONG ISLAND, NEW YORK

Fig. 8.2 Variations of soil temperature at indicated depths (ft) (Long Island, New York)

Assuming the temperature variation is sinusoidal with a period of one day, the rms systematic wander per day can be calculated from (8.4) and (8.9), yielding

$$\sigma_{il}^S[N] = \left[\frac{l}{c} \frac{\partial n_c}{\partial \theta} + \frac{n_c}{c} \frac{dl}{d\theta} \right] \frac{\theta_{p\text{-}p}}{2\sqrt{2}} \qquad (8.14)$$

where $\theta_{p\text{-}p}$ is the peak-to-peak temperature change in one day. Because we are assuming sinusoidal wander, the peak-to-peak wander per day is $2\sqrt{2}$ times the rms wander. Substituting the appropriate parameters found in Table 8.1 into (8.14) and (8.8), a 1.0°C ground temperature change along a 1000 km transmission path containing 20 fiber sections of 50 km each would result in about 40 ns of accumulated peak-to-peak diurnal wander per day, or 40ns/T UI of wander per day, assuming the temperature change is systematic along the entire route. Similarly, over the period of a year, if the ground temperature changes by 20°C, about 800 ns (800ns/T UI) of peak-to-peak annual wander would result.

With a regenerator spacing of 50 km, about 20 regenerators would be used on a 1000 km transmission path. The laser temperature in most high-speed optical regenerators is controlled by a *thermoelectric* (TE) cooler. Under normal circumstances, a TE cooler keeps the laser temperature constant to within about 2°C.

Two wavelength regions are of interest—the 1.3 μm region, where most current systems operate, and the 1.5 μm region, where some systems now operate and where many future systems will also operate. Using the appropriate parameters in Table 8.1 in (8.5), (8.11) and (8.12), and assuming the laser temperature change is sinusoidal and uncorrelated between regenerator spans, about 0.8 ns (0.8ns/T UI) of wander would result at 1.5 μm from a 2°C change in laser temperature. Much less wander would result at 1.3 μm, because of the near-zero dispersion at this wavelength.

8.3 WANDER MEASUREMENT

Wander can be measured on a fiber optic transmission path by observing the changes in propagation delay over a period of time. This can be done simply with the measurement technique described here. Other possible measurement techniques exist; this one has the advantages of simplicity and ease of implementation.

The experimental apparatus is shown in Figure 8.3. A clock with frequency stability of at least one part in 10^9 per day is used to clock a pseudorandom data pattern generator at the baud of the fiber optic transmission system. The data pattern is a known, periodic, $2^{15} - 1$ bit pattern. This electrical data pattern is split and input to an oscilloscope as well as to the first regenerator on the optical fiber path under test. The electrical output from the final regenerator is also displayed on the oscilloscope.

The oscilloscope is used to compare the input bit pattern to the output bit pattern. Wander is manifested as a movement of the output bit stream relative to the input bit stream. To measure the wander precisely over a period of time, the display is recorded at least every hour and the variation in propagation delay is noted. Two sample photographs showing the variation observed are displayed in Figure 8.4.

The results of a measurement made using this technique, are shown in Figure 8.5. The peak-to-peak wander that occurred, during the 48-hour period observed was found to be 0.950 ± 0.025 ns over a 320 km optical fiber transmission path. The pulse period was T = 2.4 ns, (1/T = 417 Mb/s) therefore about 0.4 UI peak-to-peak wander was observed in the 48-hour measurement; 0.4 UI (144 degrees) of peak-to-peak wander is significantly greater than the peak-to-peak jitter expected to be generated by a chain of six cascaded regenerators. Therefore, wander considerations should be used in determining the elastic store sizes needed at system interfaces. The measurement uncertainty results from the finite resolution of the oscilloscope screen, because according to the manufacturer's specifications, the largest possible drift in clock frequency that could occur in a day would cause only 0.003 ns of variation in the relative delay.

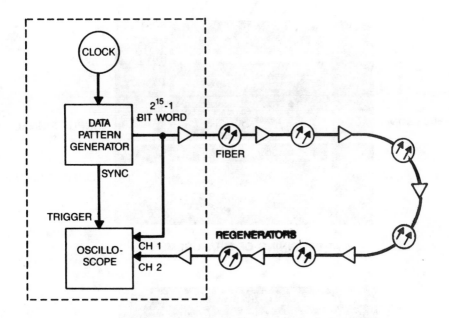

Fig. 8.3 Wander measurement experimental apparatus

Also plotted on Figure 8.5 is the air temperature obtained from the National Weather Service. Information about the air temperature was obtained to investigate the possible correlation between change in air temperature (presumably leading to a change in ground temperature) and wander. Although the peaks in air temperature roughly coincide with the overall peaks in wander, no statistical correlation can be established from this data.

The measurement must be performed over an extended period of time, at least several cycles, to determine the period of the wander. From Figure 8.5, there appear to be two major frequency components. The period of the pronounced higher frequency component cannot be determined from this data, because the variation is occurring more quickly than the measurements were made (every hour or half hour). The reason for this higher frequency variation could be fluctuations in laser wavelength, caused by small variations in laser temperature, which would result in wander. The period of the lower frequency component appears to be about 22 hours, close to the expected 24-hour period, but the poorly defined diurnal variation makes it difficult to establish the period precisely with only two cycles of data. However, it is likely that this component is caused by changes in ground temperature.

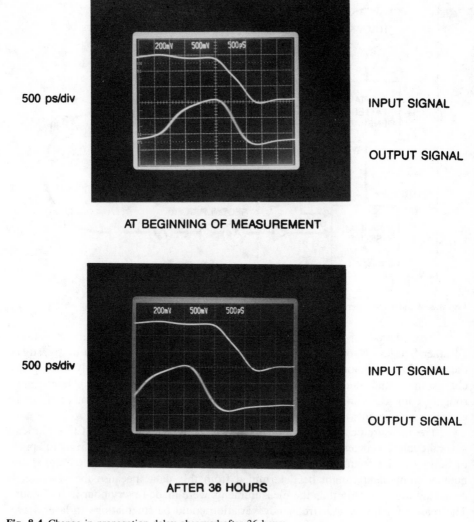

500 ps/div   INPUT SIGNAL

OUTPUT SIGNAL

AT BEGINNING OF MEASUREMENT

500 ps/div   INPUT SIGNAL

OUTPUT SIGNAL

AFTER 36 HOURS

Fig. 8.4 Change in propagation delay observed after 36 hours

8.4 REMARKS

In this chapter we have discussed wander, pulse position modulation that occurs with extremely low frequencies as a result of variations in the light propagation properties of the optical fiber itself. A model for wander was presented, including the effects of fluctuations in optical fiber temperature, laser temperature, and the reference timing signal. A simple model for wander accumulation can be developed because the

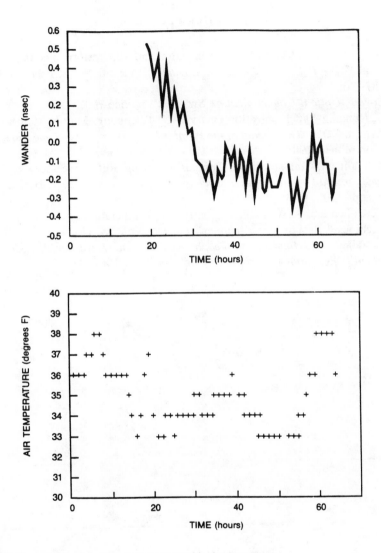

Fig. 8.5 Comparison of wander measurement with air temperature

wander transfer function is unity. Rms systematic wander was shown to accumulate more rapidly than rms random wander of the same magnitude. Finally, a wander measurement technique was described and measurement results were given.

REFERENCES

[8.1] R. Smith and L. J. Millot, "Synchronization and Slip Performance in a Digital Network," *The Radio and Electronic Engineer*, Vol. 54, No. 2, February 1984, pp. 87–96.

[8.2] J. Gowar, *Optical Communication Systems*, Prentice-Hall, 1984.

[8.3] N. J. Ronan, "Synchronization Overview," *Telephony*, May 5, 1986.

[8.4] T1X1.6/88–016, *The Impact of the Pointer Process on Phase-Time Variation*, T1 Committee, 1988.

[8.5] I. H. Malitson, "Interspecimen Comparison of the Refractive Index of Fused Silica," *Journal of the Optical Society of America*, Vol. 55, No. 10, Nov. 1965, pp. 1205–9.

[8.6] L. G. Cohen and J. W. Fleming, "Effect of Temperature on Transmission in Lightguides," *Bell System Technical Journal*, Vol. 58, No. 4, April 1979.

[8.7] S. L. Valley, ed., *Handbook of Geophysics and Space Environments, Air Force Cambridge Research Laboratories*, McGraw-Hill, 1965.

Chapter 9
Network Jitter Standards

As indicated in earlier chapters, jitter accumulates in digital networks and may degrade transmission performance. As a result of timing signals being displaced from their optimum positions in time, errors may be introduced into digital signals at points of signal regeneration. Slips may be introduced into digital signals resulting from either data overflow or depletion in digital equipment incorporating elastic stores. In addition, phase modulation of the reconstructed samples in digital-to-analog conversion devices may result in degradation of digitally encoded analog signals. To avoid the above impairments, it is necessary to control jitter accumulation in digital networks. This chapter reviews the current status of network jitter standards and control strategies for 1.544 Mb/s hierarchies within worldwide standards arenas. The discussion focuses on network jitter standards developed within the CCITT and Technical Subcommittee T1X1.

The CCITT (International Telegraph and Telephone Consultative Committee) is chartered "to study and issue recommendations on technical, operating, and tariff questions relating to telegraphy and telephony." The CCITT term "recommendation" is equivalent to the term "standard" used by other organizations. The CCITT is divided into fifteen Study Groups, each of which is assigned a specific responsibility, and operates on four-year study periods. During this time each Study Group works to answer a series of technical questions assigned by the CCITT Plenary Assembly. At the conclusion of a study period, the results of work of the Study Groups are presented to the Plenary Assembly for adoption as Recommendations, categorized in terms of alphabetically denoted Series. The approved Recommendations for each four-year study period are published in a collection of books that are identified by the color of their covers. The 1984 results are called the Red Books, and the forthcoming 1988 results will be called the Blue Books. Within this chapter we will discuss CCITT Recommendation G.824, which addresses jitter and wander[1] control for 1.544 Mb/s based hierarchies. This Recommendation falls within the scope of Question 15 of Study Group XVIII, which deals with performance objectives for timing and controlled slips (synchronization), jitter and wander, and propagation delay.

Committee T1 has become the largest affiliated committee of the American National Standards Institute (ANSI). The mission of T1 is provide the telecommunications industry with an open public forum for developing essential post-divestiture interconnection, interoperability and performance standards. These standards are applicable to US networks which form part of the North American telecommunications system. Committee T1 consists of a main committee, an elected Advisory Group, and six *Technical Subcommittees* (TSCs). Each subcommittee establishes working groups, as needed, to address specific projects. We shall describe network jitter specifications which have been developed within Technical Subcommittee T1X1, which addresses matters pertaining to the digital hierarchy and synchronization. Note that it is not within the scope of T1X1 to develop equipment specifications.

9.1 CCITT RECOMMENDATION G.824

The approach taken towards jitter control within CCITT Recommendation G.824, entitled "The Control of Jitter and Wander Within Digital Networks Which are Based on the 1544 Kbit/s Hierarchy," as revised during the 1984 study period [9.1], exhibits a high degree of commonality with the approach taken within Technical Subcommittee T1X1 [9.2] The revised Recommendation, which will appear within the forthcoming Blue Books,[2] is comprised of four sections. These sections include a control strategy, network specifications, a framework for equipment specifications, and discussion of accumulation within digital networks. Suggestions for the measurement of jitter and wander will appear as Supplement No. 3.8 of the O-Series [9.3] and Supplement No. 35 of the G-Series Recommendations, respectively, in the forthcoming Blue Books.

The objective of the control strategy detailed in revised Recommendation G.824 is to minimize impairments due to jitter and wander in digital networks. A further objective is to provide sufficient guidance so that components from different administrations and suppliers can work together and jointly meet network requirements. In order to accomplish these objectives, network specifications at hierarchical interfaces are coupled with individual equipment specifications. The strategy involves the following elements:

1. Specifications of network limits not to be exceeded at any hierarchical interface.
2. Specifications of individual digital equipment within a consistent framework.
3. Information and guidance concerning issues related to network and individual digital equipment jitter and wander performance, which
   - enables analysis and prediction of jitter and wander accumulation in arbitrary network configurations,
   - facilitates satisfactory control of impairments due to this accumulation,
   - provides insight into the jitter and wander performance (e.g., tolerance, transfer, generation) of individual digital equipments.

4. Provision of a measurement methodology to facilitate accurate and repeatable jitter and wander measurements.

The network specifications represent maximum permissible output jitter limits at hierarchical interfaces of a digital network. Specifications of jitter at hierarchical interfaces represents the next step beyond individual equipment specifications in developing a unified approach towards network jitter control. The introduction of common standards across all equipment designs is intended to further the development of equipment jitter generation specifications such that nationwide networks can generally operate to deliver jitter within the interface specifications. These specifications are contained in Table 9.1.

Table 9.1 Maximum permissible output jitter at hierarchical interfaces

Digital rate (kb/s)	Network Limit (UI peak-to-peak)		Bandpass filter having a lower cut-off frequency f_1 or f_3 and a minimum upper cut-off frequency f_4		
	B_1	B_2	f_1 (Hz)	f_3 (kHz)	f_4 (kHz)
1544	5.0	0.1[a]	10	8	40
6312	3.0	0.1[a]	10	3	60
32064	2.0	0.1[a]	10	8	400
44736	5.0	0.1	10	30	400
97728	1.0	0.05	10	240	1000

Notes:
a. This value requires further study.
For systems in which the output signal is controlled by an autonomous clock (e.g., quartz oscillator) more stringent output jitter values may be defined in the relevant equipment specifications (e.g., for the muldex in Recommendation G.743, output jitter should not exceed 0.01 UI rms).

The framework for equipment specifications provides minimum guidelines for the jitter and wander tolerance of input ports, and cross-references relevant CCITT equipment specifications for further guidance regarding tolerance, transfer, and generation requirements. It has been generally recognized that jitter transfer and generation specifications for individual digital equipment may be expected to differ. However, it has long been an objective within standards organizations to develop a single input jitter tolerance specification which could be applied to an arbitrary digital equipment to promote flexible networking. Ideally, compliance with such a specification would ensure satisfactory performance of all classes of digital equipment in the presence of network jitter. It has not yet been possible to develop this type of specification because: (1) sinusoidal jitter is typically utilized for testing (since it is easily verified with conventional test equipment), (2) the difference in the tolerance of line regenerators and bit justification multiplexes, for example, to input sinusoidal jitter is quite substantial in some instances, and (3) the test conditions are not, in themselves, intended to be representative of the type of jitter found in practice within a network. Within revised CCITT Recommendation G.824, the intent of providing a single minimum tolerance specification is to serve as a "benchmark" for equipment designers so that new equipment designs provide at least a baseline level of jitter performance. Compliance with this specification is intended to ensure that a certain minimum tolerance criteria has been satisfied, which constitutes a necessary, but not sufficient, condition for determining satisfactory equipment performance in the presence of network jitter. Therefore, it is recognized that reference to individual equipment specifications must always be made. Note that all of the tolerance requirements implicitly assume use of an "onset of errors" measurement technique (an example of which is described in Chapter 6). As discussed in Chapters 4 and 6, and noted in revised Recommendation G.824, measurements carried out to verify compliance with specifications in the region governed by the alignment jitter handling capability of the input timing extraction circuit may provide environment dependent results, allowing some ambiguity in their interpretation. The Recommendation advises that account be taken of the requirement at the design stage of the equipment. The minimum tolerance specifications are contained in Table 9.2. The parameter values for the DS1 to DS3 hierarchical levels contained in Table 9.2 are based upon the lower bound of a consistent superposition of 1.544 Mb/s based hierarchy[3] regenerator and bit justification multiplex sinusoidal input jitter tolerance requirements and characteristics. The parameter values chosen account for the range of existing allowable regenerator timing extraction circuit corner frequencies [9.4].

Cross-referenced jitter specifications include those for positive bit justification multiplexes, contained in Recommendations G.743 and G.752. These Recommendations address the jitter tolerance, transfer, and generation requirements for DS1 to DS2 and DS2 to DS3 multiplexes, respectively. As an example, we will discuss the input jitter tolerance requirements for a DS1 to DS2 multiplexer (shown in Figure 9.1). CCITT requirements for maximum tolerable input jitter make use of a mask with six specified parameters: A_1, A_2, f_1, f_2, f_3, and f_4. These parameters are based upon the following considerations:

- The parameter A_1 refers to the maximum peak-to-peak amplitude of jitter that should be accommodated.
- The parameter A_2 is intended to reflect the alignment jitter handling capability of the input timing extraction circuit. This may be thought of in terms of the percentage of the timing extraction circuit "eye" allocation to alignment jitter.
- The parameter f_1 represents a low frequency limiting value, and serves as a demarcation of the defined jitter/wander frequency regions. This parameter also reflects practical consideration regarding generation and measurement of large sinusoidal jitter amplitudes at low frequencies.

Table 9.2 Jitter and wander tolerance of input ports (Provisional values) (Note 1)

Bit Rates (kb/s)	Jitter amplitude (peak-to-peak)			Frequency					Test Signal
	A_0 (μs)	A_1 (UI)	A_2 (UI)	f_0 (Hz)	f_1 (Hz)	f_2 (Hz)	f_3 (kHz)	f_4 (kHz)	
1544	18 (Note 2)	5.0	0.1 (Note 2)	1.2×10^{-5}	10	120	6	40	$2^{20} - 1$ (Note 3)
6312	18 (Note 2)	5.0	0.1	1.2×10^{-5}	10	50	2.5	60	$2^{20} - 1$ (Note 2)
32064	18 (Note 2)	2.0	0.1	1.2×10^{-5}	10	400	8	400	$2^{20} - 1$ (Note 2)
44736	18 (Note 2)	5.0	0.1 (Note 2)	1.2×10^{-5}	10	600	30	400	$2^{20} - 1$ (Note 2)
97728	18 (Note 2)	2.0	0.1	1.2×10^{-5}	10	12000	240	1000	$2^{23} - 1$ (Note 2)

Notes:
1. Reference to individual equipment specifications should always be made to check if supplementary input jitter tolerance requirements are necessary.
2. This value requires further study.
3. It is necessary to suppress long zero strings in the test sequence in networks not supporting 64 kbit/s transparency.
4. The value A_0 (18 μs) represents a relative phase deviation between the incoming signal and the internal local timing signal derived from the reference clock.

- The parameter f_3 refers to the effective "Q-factor" of the clock extraction circuits used at the multiplexer input ports.
- The value of the parameter f_2 is obtained by extending the (f_3, A_2) coordinate at a slope of 20 dB/decade until it intersects the peak-to-peak amplitude A_1.
- The parameter f_4 represents a high frequency limiting value. This parameter reflects practical considerations regarding generation and measurement of low amplitude sinusoidal jitter at high frequencies, which may be limited by the capabilities of readily available jitter generation and measurement equipment. Therefore, an upper frequency limit of $f_4 = 5f_3$ was chosen.

Comparing Figure 9.1 and Figure 6.1, we see that the ability of a multiplex to accommodate sinusoidal jitter of peak-to-peak amplitude A_1 and frequency f_2 is related to the justification capacity and synchronizer elastic store size (Region 3 of Figure 6.1), whereas its ability to accommodate sinusoidal jitter of peak-to-peak amplitude A_2 and frequency f_3 is determined by the input timing extraction circuit (Region 4 of Figure 6.1). Using the notation of Chapter 6, $A_1 = 2$ UI peak-to-peak corresponds

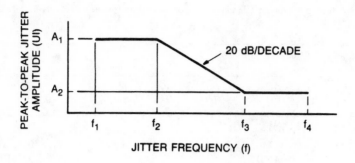

INPUT	A₁ (UI)	A₂ (UI)	f₁ (Hz)	f₂ (Hz)	f₃ (kHz)	f₄ (kHz)
1544 kbit/s	2	0.05	10	200	8	40

Fig. 9.1 CCITT G.743 DS1 to DS2 multiplexer input tolerance requirements

to $A_I = 1$ UI peak, and $f_I = 200$ Hz. Applying (6.17), the magnitude of the peak frequency deviation from the nominal is $2\pi A_I f_I = 1257$ Hz. As shown in Chapter 6, the lowest value of the maximum tolerable frequency deviation from the nominal which satisfies the requirement for normal justification for a DS1 to DS2 multiplexer is $f_{P-} = 1545$ Hz. As 1257 Hz is less than f_{P-}, we see that Recommendation G.743 places no constraints on synchronizer elastic store size for a DS1 to DS2 multiplexer.

Referring to the discussion of Chapter 4, the 20 dB/decade slope in Figure 9.1 reflects the sinusoidal input tolerance characteristic of a first-order timing extraction circuit, f_3 reflects its corner frequency, and A_2 is intended to reflect the proportion of the regenerator "eye" allocated to input jitter. [9.5] indicates that the value for the parameter A_2 originated as a result of tests to determine crosstalk margins and minimum operational cross-connect signal levels.

It should be noted that analogous bit justification multiplexer input tolerance specifications generally used within the U.S. (shown in Figure 9.2) [9.6,9.7] are more stringent. These specifications also make use of a mask with the six specified parameters discussed above. However, the basis for the parameters A_1 and f_2 differ:

- The parameter A_1 is close in value to that of the number of cells in the synchronizer elastic store designated for network input jitter.
- The parameter f_2 is chosen such that the maximum frequency deviation $(2\pi)A_1/2(f_2)$ may be accommodated only when the designated number of synchronizer elastic store cells have been allocated.
- The template is formed by connecting the coordinates (A_1,f_1), (A_1,f_2), (A_2,f_3), and (A_2,f_4).

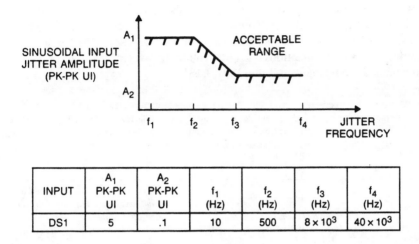

INPUT	A_1 PK-PK UI	A_2 PK-PK UI	f_1 (Hz)	f_2 (Hz)	f_3 (Hz)	f_4 (Hz)
DS1	5	.1	10	500	8×10^3	40×10^3

Fig. 9.2 U.S. DS1 to DS2 multiplexer input tolerance requirements

Though these specifications were derived for a multiplexer in isolation, they are generally applied equally to a multiplexer-demultiplexer pair. As shown in Chapter 6, application of these specifications to a DS1 to DS2 multiplexer requires at least five cells allocation to input jitter in the synchronizer and desynchronizer elastic stores.

The jitter accumulation models contained in Supplement No. 36 of the G-Series Recommendations are consistent with those described in Chapter 7.

9.2 T1X1 NETWORK JITTER SPECIFICATIONS

An ongoing work effort within Technical Subcommittee T1X1 has been focused towards the development of a strategy for the control of jitter and wander in digital networks based on the 1.544 Mb/s hierarchy. This strategy, and associated T1X1 documentation, is intended to provide:

1. A jitter and wander control framework to motivate and outline the overall control strategy, indicating the primary issues and areas which need to be addressed [9.2].
2. Characterization guidelines to provide information and guidance concerning issues relating to jitter and wander performance [9.7,9.8].
3. Standardized measurement methodology to facilitate accurate and repeatable jitter and wander measurements for complex networks as well as individual digital equipment [9.9,9.10].
4. Network interface jitter and wander specifications [9.11,9.12].
5. Appropriate reference to equipment specifications.

We should note that work in all of the above categories is reflected in the forthcoming CCITT Blue Books. Network output jitter specifications at hierarchical interfaces will be included in the next issue of T1.102-1987, "American National Standard for Telecommunications—Digital Hierarchy—Electrical Interfaces." These specifications are contained in Table 9.3 [9.13].

Table 9.3 Limits for Maximum Permissible Network Output Jitter at DSX-1, DSX-1C, DSX-2, and DSX-3 Interfaces. (These limits represent the maximum permissible levels of output jitter at DSX-1, DSX-1C, DSX-2, and DSX-3 interfaces within a digital network. It is emphasized that these limits include jitter accumulated from transmission source to sink.)

	Network Limit Peak-to-Peak UI		Measurement Filter Bandwidth Bandpass Filter Having a Lower Cut-off Frequency F_1 or F_3 and a Minimum Upper Cut-off Frequency F_4		
Level	B_1 $(F_1 - F_4)$	B_2 $(F_3 - F_4)$	F_1 (Hz)	$F_3{}^1$ (kHz)	F_4 (kHz)
DS1	5.0	0.1	10	8	40
DS1C	5.0	0.1	10	1.5	40
DS2	3.0	0.1	10	3	60
DS3	5.0	0.1	10	30	400

Notes:
1. 1.0 DS1 UI = 648 ns; 1.0 DS1C UI = 317 ns; 1.0 DS2 UI = 158 ns; 1.0 DS3 UI = 22.4 ns.
2. The frequency response of the measurement filters should have a roll-off of $\pm$ 20 dB/decade.
3. The measurement time interval requires further study; it is provisionally recommended that the measurement time interval be at least one minute.
4. The frequency F_3 represents the jitter half-bandwidth of (typical) timing extraction circuits. The frequency F_3 is given by $F_3 = f_0/(2Q)$, where f_0 and Q represent the line rate and Q-factor of the timing extraction circuit, respectively.

9.3 REMARKS

In this chapter we have examined and compared CCITT and T1X1 approaches to network jitter standards and control strategies. Substantial progress has been achieved over this study period towards convergent worldwide specifications in this area.

NOTES

1. Standards arenas currently provide separate specifications for jitter and wander; for 1.544 Mb/s based hierarchies jitter is defined as limited to phase variations

above 10 Hz and wander to those below 10 Hz.
2. Assuming it is approved by the Plenary Assembly.
3. The United States, Canada, Korea, Taiwan, and Japan are examples of countries which use levels of this hierarchy.

REFERENCES

[9.1] CCITT Report R65, Doc. APIX-150-E, *Final Report to the IXth CCITT Plenary—Part X*, July 1988.

[9.2] T1X1.4/86-015, "Strategy for the Control of Jitter and Wander Within Digital Networks Which are Based on the 1.544 Mbps Hierarchy," *T1 Committee*, 1986.

[9.3] CCITT Study Group SVIII, Geneva, 6–17 June 1988, Question: 15/XVIII, 10IV, "Supplement 3.8-Guidelines Concerning the Measurement of Jitter as Agreed by Study Group IV," TD 25.

[9.4] "The Low Power T1 Line Repeater Compatibility Specification," *AT&T Compatibility Bulletin* No 113, Issue 2, April 1978.

[9.5] CAN COM XVII-No. 1, COM XVII, "Jitter Tolerances for Digital Multiplex Equipment According to Recommendations G.743 and G.752," July 1978.

[9.6] "Transport Systems Generic Requirements (TSGR): Common Requirements," *Bellcore Technical Reference* TR-TSY-000499, Issue 1, December 1987.

[9.7] "Digital Multiplexes, Requirements and Objectives," *Bell System Technical Reference*, PUB 43802, July 1982.

[9.8] T1X1.4/86-029, "Characterization Guidelines: Desynchronizer Jitter Transfer," *T1 Committee*, 1986.

[9.9] T1X1.3/86-026, "A Model For Computing Jitter and Wander Accumulation in Digital Networks Arising from Cascaded Digital Regenerators and Asynchronous Multiplexes," *T1 Committee*, 1986.

[9.10] T1X1.4/87-024, "Jitter Measurement Methodology," *T1 Committee*, 1987.

[9.11] T1X1.3/88-006, "Slave Stratum Clock Performance Measurement Guidelines," *T1 Committee*, 1988.

[9.12] T1X1.4/87-025, "Limits for the Maximum Permissible Network Output Jitter at any Hierarchical Interface," *T1 Committee*, 1987.

[9.13] ANSI T1.101-1986, "American National Standard—Synchronization Interface Standards For Digital Networks."

[9.14] T1X1.6/88-010, Edited Version of T1X1.4/87-025, *T1 Committee*, 1988.

Glossary of Symbols, Acronyms and Abbreviations

Symbol	Definition
a_n	binary message sequence (0,1)
a'_n	regenerated message sequence
A_I	amplitude of sinusoidal input jitter
A_N	amplitude of the extracted timing signal
AGC	automatic gain control
ANSI	American National Standards Institute
APD	avalanche photodiode
α_1	the portion of pattern dependent jitter generated by the $1th$ regenerator that accumulates systematically
b_m	the expected value of $a_n a_{n+m} e_{N-1}[nT]$, $E\{a_n a_{n+m} e_{N-1}[nT]\}$
B	bandwidth (corner frequency)
BER	bit error rate
c	speed of light, 3×10^8 m/sec
$C_I(t)$	covariance of $e_I(t)$
$C_S(t)$	covariance of $e_S(t)$
CCITT	International Telegraph and Telephone Consultative Committee
D	the timing signal is frequency divided by D to extend the range of jitter detection
D_o	linear frequency drift of reference clock
DC	zero frequency
DFB	distributed feedback, a type of semiconductor laser
$\delta(t)$	delta function, 1 for $t = 0$, 0 otherwise
e	electron charge, 1.6×10^{-19} ampere-sec
$e_{pm}(t)$	phase noise of the reference clock
$e_{fm}(t)$	frequency modulation of the reference clock
$e[nT]$	description of jitter as a discrete time sequence
$e_{aN}[nT]$	the alignment jitter at the Nth cascaded regenerator
e_{aN}^{P-P}	the peak-to-peak alignment jitter at the Nth cascaded regenerator

$e_{aN}^{R}[nT]$	the random alignment jitter at the Nth cascaded regenerator
$e_{aN}^{S}[nT]$	the systematic alignment jitter at the Nth cascaded regenerator
$e_D(t)$	the jitter on the demultiplexer output signal
$e_{iN}[nT]$	the jitter generated internally by the Nth cascaded regenerator
$e_{iN}^{R}[nT]$	the random jitter generated internally by the Nth cascaded regenerator
$e_{iN}^{S}[nT]$	the systematic jitter generated internally by the Nth cascaded regenerator
$e_{iN}^{P}[nT]$	the pattern-dependent, noise-independent jitter generated internally by the Nth cascaded regenerator
$e_{iN}^{\eta P}[nT]$	the noise-dependent, pattern-dependent jitter generated internally by the Nth cascaded regenerator
$e_{iN}^{\eta}[nT]$	the noise-dependent jitter generated by the Nth cascaded regenerator
$e_{rel}(t)$	relative peak jitter
$e_N[nT]$	Nth regenerator's output jitter
$e_N^{R}[nT]$	Nth regenerator's output random jitter
$e_N^{S}[nT]$	Nth regenerator's output systematic jitter
$e_N^{(L)}[nT]$	Nth regenerator's output jitter after data passes around the loop L times
$e_{\text{SPC}}[nT]$	synchronizer phase comparator output for a bit written into the synchronizer elastic store at time $n\,T_{nom}$
$e_I(t)$	input jitter to multiplexer
$\dot{e}_I(t)$	frequency deviation caused by input jitters
$e_{oh}(t)$	overhead bit jitter
$e_s(t)$	jitter on the gapped synchronizer read clock output
$e_{s1}(t)$	sample and hold component of $e_s(t)$
$e_{s2}(t)$	saw-toothed component of $e_s(t)$
$e_j(t)$	saw-toothed justification jitter waveform
$e_{\text{SPC}}(t)$	synchronizer phase comparator output
$e_{\text{WT}}(t)$	waiting time jitter
$erfc(x)$	complimentary error function
$E[x]$	expected value of x
$\eta_N(t)$	receiver noise of Nth regenerator
η	bandpass filter mistuning factor
ε_d	distributive error
ε_q	quantization error
ε	distributive plus quantization error
ε	element of
f	frequency
$f(t)$	transfer function of the fiber span in Chapter 1
$f(t)$	line frequency

f_c	center frequency of timing circuit bandpass filter
f_{in}	frequency of multiplexer input signal
$f_{j,max}$	maximum justification rate
$f_{j,nom}$	nominal justification rate
$f_{net}(t)$	rate of phase accumulation in synchronizer elastic store
f_k	frequency spacing of discrete components of $\bar{S}_{rr}(f)$
f_r	synchronizer read clock frequency
$f_N(t)$	the pulse shape of an isolated "*1*" to be transmitted by the *Nth* regenerator
f_{pm}	frequency of sinusoidual jitter used for testing
f_N	maximum tolerable negative frequency deviation
f_{nom}	nominal frequency of multiplexer input signal
f_{out}	multiplexer output frequency
f_P	maximum tolerable positive frequency deviation
FWHM	full width, half maximum
$g(t)$	impulse response of phase-smoothing circuit
$g_N(t)$	the pulse shape of an isolated "1" received by the *Nth* regenerator
G_{APD}	gain of APD receiver
$G(f)$	Fourier transform of $g(t)$
G_{max}	maximum $|G(f)|$
$G_N(f)$	Fourier transform of $g_N(t)$
h	Planck's constant, 6.62×10^{-34} W/sec^2
$h_N[kT]$	*Nth* regenerator's jitter impulse response
$H_N(f)$	*Nth* regenerator's jitter transfer function
$H_a(f)$	a jitter transfer function with no jitter peaking
$H_b(f)$	a jitter transfer function with 0.2 dB jitter peaking
$i(t)$	received photo current
j	$\sqrt{-1}$
J_k	Bessel function of the first kind
K	number of bits circulating in loop simulation, number of bits past and future considered in P_E calculation
K_i	the magnitude of input sinusoidal jitter used for testing
$K_o(f_{pm})$	magnitude of output sinusoidal jitter
K_p	phase detector constant in millivolts/degree
l_o	optical power out of the laser when no pulse is transmitted
l	length of fiber
$l(t)$	the light signal power from the laser transmitter
λ	wavelength
m	number of per tributary bits between overhead bits
$m(t)$	the optical pulse power
m_{fm}	frequency modulation index
M	number of time-division multiplexed low speed tributaries

M_D	desynchronizer elastic store size
M_S	synchronizer elastic size
MTBI	meantime between impairments
n_c	the fiber core's refractive index
n_{oh}	number of contiguous multiplexers overhead bits
N	number of cascaded regenerators
N_{data}	total number of data bits in group of m bits
N_{oh}	total number of overhead bits in group of m bits
$\bar{N}(x)$	average number of times per second that a threshold $|x|$ is exceeded
N_0	noise power spectral density
NRZ	non-return-to-zero (coding)
$p(e_{aN}[nT])$	probability density of alignment jitter
$p_N(t)$	the Nth regenerator's prefilter impulse response
P_D	demultiplexer output jitter power
ΔP_D	power added by justification
P_E	probability of error
P_{Ei}	probability of error of the ith message sequence
P_I	input jitter power
$P(x)$	probability of x
$P_{out,n}$	outgoing pointer adjustment from nth node
P_n	probability of exceeding the threshold $|x|$ more than n times in the time interval $(t, t + \Delta t)$
ρ	nominal justification ratio
$\rho(t)$	instantaneous justification ratio
ϕ_P	slope of the argument of $H_N(f)$ in deg/kHz, i.e., $\dfrac{d}{df}\theta(f)\big|_{f=0}$
$\Delta\Phi$	relative wander between read and write clocks in pointer processing elastic store
$\Phi_{in,n}$	wander affecting write clock phase for nth node
$\Phi_{out,n}$	wander affecting read clock phase for nth node
$\Phi_N(f)$	power spectrum of the accumulated jitter at the output of the Nth regenerator
$\Phi_I(f)$	power spectrum of input jitter
$\Phi_D(f)$	power spectrum of demultiplexer output jitter
$\Phi_N^R(f)$	power spectrum of the accumulated random jitter
$\Phi_{aN}^R(f)$	power spectrum of the accumulated random alignment jitter
$\Phi_{aN}^S(f)$	power spectrum of the accumulated systematic alignment jitter
$\Phi_N^S(f)$	power spectrum of the accumulated systematic jitter
Φ_{iN}^S	power spectral density of the systematic jitter generated by the Nth regenerator

Φ_{iN}^{R}	power spectral density of the random jitter generated by the *Nth* regenerator
$\Phi_d(f)$	power spectrum of multiplex distortion terms
$\Phi_{f-t}(f)$	power spectrum of multiplex feed-through terms
Φ_{il}^{PD}	power spectrum density of the pattern dependent jitter at the *Nth* regenerator
Φ_{iN}^{P}	power spectral density of the pattern dependent, noise independent jitter generated by the *Nth* regenerator
$\Phi_{iN}^{\eta P}$	power spectral density of the pattern dependent, noise dependent jitter generated by the *Nth* regenerator
Φ_{iN}^{η}	power spectral density of the noise dependent jitter generated by the *Nth* regenerator
$\Phi_S(f)$	power spectrum of $e_S(t)$
$\Phi_j(f)$	power spectrum of $e_j(t)$
PPL	phase locked loop
$Q[x]$	quantized version of signal x
q	quantum efficiency of photodiode
$q_N(t)$	pulse shape of an isolated "*l*" after the *Nth* regenerator's prefilter
$Q_N(f)$	the Fourier transform of $q_N(t)$
$r_N(t)$	the *Nth* regenerator's received signal
$r_N'(t)$	the *Nth* regenerator's output signal
rms	root mean squared
RLC	resistor, inductor, capacitor circuit
RZ	return-to-zero coding
$R_{aa}[m]$	the autocorrelation of a_N
$R_j(\tau)$	the autocorrelation of $e_j(t)$
$R_{qq}(\tau)$	the autocorrelation of $q_N(t)$
$R_{q^2q}^2(t)$	the autocorrelation of $q_N^2(t)$
$R_{rr}(\tau)$	the autocorrelation of $r_N(t)$
$R_{yy}(\tau)$	the autocorrelation of $y_N(t)$
$s_N(t)$	*Nth* regenerator's timing signal
$S_{aa}(f)$	power spectrum of $a_n(t)$
$S_{rr}(f)$	power spectrum of $r_N(t)$
$S_{yy}(f)$	power spectrum of $y_N(t)$
$S_{q^2q^2}(f)$	power spectrum of $q_N^2(t)$
step$(x - a)$	step function, 1 if $x > a$, 0 if $x < a$
SAW	surface acoustic wave
SNR	signal-to-noise ratio
SDH	synchronous digital hierarchy
SONET	synchronous optical network
STS-N	synchronous transport signal level N

$\sigma^L[N]$	standard deviation of the jitter after data passes around the loop L times
σ^D	rms demultiplexer output jitter
$\sigma_R^{99\%}[N]$	99% confidence bounds of random rms jitter
$\sigma_S^{99\%}[N]$	99% confidence bounds of systematic rms jitter
$\sigma[N]$	rms jitter at the output of the Nth regenerator
$\sigma^{\eta P}[N]$	rms noise dependent, pattern dependent jitter of the Nth regenerator
$\sigma^P[N]$	rms noise independent, pattern dependent jitter of the Nth regenerator
$\sigma_R[N]$	random rms jitter at the output of the Nth regenerator
$\sigma_S[N]$	systematic rms jitter at the output of the Nth regenerator
σ_{iN}^S	systematic rms jitter generated by the Nth regenerator
σ_{iN}^R	systematic rms jitter generated by the Nth regenerator
$\sigma_{aR}[N]$	rms random alignment jitter at the output of the Nth regenerator
$\sigma_{aS}[N]$	rms systematic alignment jitter at the output of the Nth regenerator
t	time
t_j	time between justification opportunities
t_n	sampling times
t_{thr}	time between ith justification opportunity and decision threshold crossing
t_0	static time shift caused by non-zero decision threshold
t_P	time $\dot{e}_l(t)$ equals f_P
t_N	time $\dot{e}_l(t)$ equals f_N
T	period of input signal
T_{nom}	nominal period of input signal
T_r	period of synchronizer read clock
TAT-8	eighth translantic telephone cable, first to be fiber optic
τ	time delay
τ_L	delay around loop
θ_1	laser temperature
θ	fiber temperature
$\Theta(f_{Pm})$	argument of the jitter transfer function
$u_N(t)$	impulse response of Nth regenerator's bandpass filter
$U_N(f)$	transfer function of Nth regenerator's bandpass filter
UI	unit interval, 360° of jitter at bit period T
v	periodic sequence length
V	sampling threshold voltage
$w[nT]$	description of wander as a discrete time function
$w_N^R[nT]$	Nth regenerator's random wander output
$w_N^S[nT]$	Nth regenerator's systematic wander output

$w_{iN}^{\theta}[nT]$	wander caused by fiber temperature changes generated by the *Nth* fiber span
$w_{iN}^{\lambda}[nT]$	wander caused by a laser wavelength change in the *Nth* regenerator
$w_{f_0}[nT]$	wander of the reference clock distribution
$x_N(t)$	*Nth* regenerator's prefilter output signal
y_b	frequency offset from reference clock disruption
$y_N(t)$	*Nth* regenerator's squares output signal
ξ	damping factor of filter
$Y_N(f)$	Fourier transform of $y_N(t)$
$*$	convolution
$\boxed{*}$	discrete convolution
∞	infinity
$[x]$	greatest integer of x
Δ	elastic storage cell delay margin
Λ	justification decision threshold
μ_s	expected value of $e_s(t)$
$\Delta_1 e_s[nT_{nom}]$	phase change from one bit to next of gapped synchronizer read clock signal
$\Delta_1 e_{SPC}[nT_{nom}]$	phase change from one bit to the next of the synchronizer phase comparator
$\Delta_m e_s[nT_{nom}]$	phase change over m bits of gapped synchronizer read clock signal
$\Delta_m e_{SPC}[nT_{nom}]$	phase change over m bits of synchronizer phase comparator for bits written into the synchronizer elastic store at times nT_{nom}

Index